Lecture Notes in Physics

For information about Vols. 1–110, please contact your bookseller or Springer-Verlag.

Lecture Notes in Physics

Edited by H. Araki, Kyoto, J. Ehlers, München, K. Hepp, Zürich
R. Kippenhahn, München, H. A. Weidenmüller, Heidelberg
and J. Zittartz, Köln

188

A. P. Balachandran · G. Marmo
B.-S. Skagerstam · A. Stern

Gauge Symmetries
and Fibre Bundles

Applications to Particle Dynamics

Springer-Verlag
Berlin Heidelberg GmbH 1983

Authors

A. P. Balachandran [*]
Physics Department, Syracuse University
Syracuse, NY 13210, USA

G. Marmo
Istituto di Fisica Teorica, Universita degli Studi Napoli
Mostra d'Oltremare Pad. 19, I-80125 Napoli, Italy

B.-S. Skagerstam [†]
Institute of Theoretical Physics, University of Gothenburg
S-41296 Gothenburg, Sweden

and

Nordita, Blegdamsvej 17, DK-2100 Copenhagen, Denmark

A. Stern [††]
Institute of Theoretical Physics, University of Gothenburg
S-41296 Gothenburg, Sweden

[*] Supported by the U.S. Department of Energy under
Contract No. De – AC02 – 76ER03533

[†] Supported by the Swedish Natural Science Research Council
under Contract No. 7310-111

[††] On leave of absence from Center for Particle Theory,
University of Texas at Austin, Austin, TX 78712, USA

ISBN 978-3-540-12724-6   ISBN 978-3-540-38707-7 (eBook)
DOI 10.1007/978-3-540-38707-7

2153/3140-543210

CONTENTS

1. Introduction

A theory defined by an action which is invariant under a time dependent group of transformations can be called a gauge theory. Well known examples of such theories are those defined by the Maxwell and Yang-Mills Lagrangians. It is widely believed nowadays that the fundamental laws of physics have to be formulated in terms of gauge theories.

The underlying mathematical structures of gauge theories are known to be geometrical in nature and the local and global features of this geometry have been studied for a long time in mathematics under the name of fibre bundles. It is now understood that the global properties of gauge theories can have a profound influence on physics. For example, instantons and monopoles are both consequences of properties of geometry in the large, and the former can lead to e.g. C P violation, while the latter can lead to such remarkable results as the creation of fermions out of bosons. Some familiarity with global differential geometry and fibre bundles seems therefore very desirable to a physicist who works with gauge theories. One of the purposes of the present work is to introduce the physicist to these disciplines using simple examples.

There exists a certain amount of literature written by general relativists and particle physicists which attempts to explain the language and techniques of fibre bundles. Generally, however, in these admirable reviews,

the concepts are illustrated by field theoretic examples
like the gravitational and the Yang-Mills systems. This
practice tends to create the impression that the subtleties
of gauge invariance can be understood only through the
medium of complicated field theories. Such an impression
however is false and simple systems with gauge invariance
occur in plentiful quantities in the mechanics of point
particles and extended objects. Further, it is often the
case that the large scale properties of geometry play an
essential role in determining the physics of these systems.
They are thus ideal to commence studies of gauge theories
from a geometrical point of view. Besides, such systems
have an intrinsic physical interest as they deal with
particles with spin, interacting charges and monopoles,
particles in Yang-Mills fields, etc. We shall present
an exposition of these systems and use them to introduce
the reader to the mathematical concepts which underlie gauge
theories. Many of these examples are known to exponents of
geometric quantization, but we suspect that due in part to
mathematical difficulties, the wide community of physicists
is not very familiar with their publications. We admit that
our own acquaintance with these publications is slight. If
we are amiss in giving proper credit, the reason is ignorance
and not deliberate intent.

The matter is organized as follows. After a brief
introduction to the concept of gauge invariance and its
relationship to determinism, we introduce in chapters 3

and 4 the notion of fibre bundles in the context of a
discussion on spinning point particles and Dirac monopoles.
The fibre bundle language provides for a singularity-
free global description of the interaction between a
magnetic monopole and an electrically charged test particle.
Chapter 3 deals with a non relativistic treatment of the
spinning particle. The non trivial extension to relativistic
spinning particles is dealt with in Chapter 5. The free
particle system as well as interactions with external
electromagnetic and gravitational fields are discussed in
detail. In chapter 5 we also elaborate on a remarkable
relationship between the charge-monopole system and the
system of a massless particle with spin. The classical
description of Yang-Mills particles with internal degrees
of freedom, such as isospin or colour, is given in chapter
6. We apply the above in a discussion of the classical
scattering of particles off a 't Hooft-Polyakov monopole.
In chapter 7 we elaborate on a Kaluza-Klein description
of particles with internal degrees of freedom. The canonical
formalism and the quantization of most of the preceeding
systems are discussed in chapter 8. The dynamical systems
given in Chapters 3-7 are formulated on group manifolds.
The procedure for obtaining the extension to super-group
manifolds is briefly discussed in chapter 9. In chapter
10, we show that if a system admits only local Lagrangians
for a configuration space Q, then under certain conditions,
it admits a global Lagrangian when Q is enlarged to a suitable U(1) bundle over Q. Conditions under which a symplectic
form is derivable from a Lagrangian are also found.

The list of references cited in the text is, of course, not complete, but it is instead intended to be a guide for the extensive literature in the field.

2. Meaning of Gauge Invariance.

Below we will deal with systems which exhibit a gauge symmetry. It is thus useful to clarify the distinction between ordinary symmetries and gauge symmetries at the beginning.

2.1 The Action

The action S is a functional of fields with values in a suitable range space. The domain of the fields is a suitable parameter space.

Thus for a nonrelativistic particle, the range space may be R^3, a point of which denotes the coordinate of the particle. The parameter space is R^1, a point of which denotes an instant of time. The fields are functions from R^1 to R^3:

$$F(R^1, R^3) = \{q\}, \quad q = (q_1, q_2, q_3), \quad q(t) \; \varepsilon \; R^3 \qquad (2.1)$$

Thus each field q assigns a point $q(t)$ in R^3 to each instant of time t.

For a real scalar field theory in Minkowski space M^4, the parameter space is M^4, the range space is R^1 and the set of fields $F(R^4, R)$ is the set of functions from R^4 to R^1.

Let us denote the parameter space by D, the range space by R and the set of fields by $F(D,R)$. Then the action S is a functional on $F(D,R)$. It assigns to each $f \; \varepsilon \; F(D,R)$ a number $S(f)$. For instance, in the nonrelativistic example cited above,

$$S(q) = \frac{m}{2} \int dt \; \frac{dq_i(t)}{dt} \frac{dq_i(t)}{dt} \qquad (2.2)$$

[The action also depends on the limits of the time integration. Since these limits are not important for us, they have here been ignored. If necessary, they can be introduced by restricting D suitably. In this case, for example, instead of R^1, we can choose for D the interval $t_1 \leq t \leq t_2$.]

The concept of a <u>global symmetry group</u> $G = \{g\}$ may be defined as follows: Suppose G is a group with a specified action $r \to gr$ on $R \equiv \{r\}$. Then, G has a natural action $f \to gf$ on $F(D,R)$, where $(gf)(t) = gf(t)$. This group of transformations on $F(D,R)$ is the global group associated with G.We denote it by the same symbol G.We say further that G is a global symmetry group if

$$S(f) = S(gf) \tag{2.3}$$

upto surface terms.For simplicity we shall assume hereafter that G is a Lie group.

As an example, consider the nonrelativistic free particle with $D = \{ t| -\infty < t < \infty \}$, $R=R^3$ and $G = SO(3)$.The rotation group has a standard action on R^3.It can be "lifted" to the action $q \to gq$ on $F(R^1,R^3)$, where

$$\big[g q\big](t) = gq(t) \qquad \big[\equiv (g_{ij}q_j(t))\big] \tag{2.4}$$

Thus in the usual language, g is a global rotation.Further, $SO(3)$ is a global symmetry group since for (2.4)

$$S(q) = S(gq) \tag{2.5}$$

In contrast, the gauge group G associated with G is defined to be the set of all functions from D to G, i.e. $G=F(D,G)=\{h\}$, and for $d \in D$, $d \overset{h}{\to} h(d) \in G$. The group multiplication in G is defined by $(hh')(d)=h(d)h'(d)$. This group, as well, has a natural action on $F(D,R)$: $(hf)(d)=h(d)f(d)$. If S is invariant under G (up to surface terms), i,e. $S(f)=S(hf)$ + possible surface terms, then the gauge group is a gauge symmetry group.

It is possible that the sort of boundary conditions we impose on the set of functions in the gauge group can have serious consequences for the theory |1|.If we do not impose any particular boundary conditions (so that the boundary conditions are "free"), G will contain constant functions and the associated global group G may be thought of as the subgroup of G of these constant functions.

Let G be a gauge symmetry group and Γ be a global symmetry group not associated with G.Recall that the parameter space contains a coordinate which we identify as time t.The profound difference bet- ween G and Γ is due to the fact that G contains time dependent symme- tries unlike Γ .It affects the deterministic aspects of the theory and also has its impact on Noether s derivation of conservation laws. These twin aspects are manifested as constraints in the Hamiltonian frame work [2] .We can illustrate these remarks as follows:

a) Determinism

A trajectory in our language is a function $\bar{f}$ ϵF (D,R) such that

$$\delta S(\bar{f}) = 0 \qquad\qquad (2.6)$$

Suppose $\bar{f}$ is a possible trajectory for a specified set of initial conditions $d^k\bar{f}/dt^k\big|_{t=0}$, k=0,1,...,n.Since G is a symmetry group, $h\bar{f}$ is also a trajectory.Further since the time dependence of h is at our disposal, we can choose h such that

$$\frac{d^k(h\bar{f})(t)}{dt^k}\bigg|_{t=0} = \frac{d^k\bar{f}(t)}{dt^k}\bigg|_{t=0} , \quad k=0,1,...,n \qquad (2.7)$$

This does not constrain h to be trivial for all time.(We assume of course that G acts non-trivially on fields.)The conclusion is that there are several possible trajectories for specified initial condi- tions.In this sense, the theory does not determine the future from the present if the state of the system is given by the values of $\bar{f}$ and its derivatives at a given time.

In the customary formulation, determinism is restored by consider- ing only those functions which are invariant under G.These gauge invariant functions and their derivatives at a given time are then defined to constitute the observables of the theory.(Such a definition of observables seems to have little direct bearing on whether they

are accesible to experimental observation. It is a definition which
is _internal_ to the theory and dictated by requirements of determinism.)

In a Hamiltonian formulation with no constraints, the specifi-
cation of Cauchy data (a point of phase space) allows us to uniquely
specify the future state of the system (at least for sufficiently
small times). The existence of a gauge symmetry group for the action
S thus means that S should lead to a constrained Hamiltonian dynamics.
An orderly way to treat such a dynamics is due to Dirac [2] .We will
have occasion to use it later.

b) Conservation Laws

The infinitesimal variation of S under a gauge transformation
is characterized by arbitrary functions ϵ_α .If G is a symmetry, Noether's argument shows that there is a charge

$$Q = \int_{\overline{D}} \epsilon_\alpha Q_\alpha \qquad (2.8)$$

which is a constant of motion

$$\frac{dQ}{dt} = 0 \qquad (2.9)$$

Here $\overline{D}$ is a fixed time slice of D. Since the ϵ_α's are arbitrary functions, we can conclude that [3]

$$Q_\alpha = 0 \qquad (2.10)$$

Thus the generators of the gauge group vanish identically.

In electromagnetism, the analogues of (2.10) are Gauss'
law

$$\nabla \cdot \vec{E} + j_o = 0 \qquad (2.11)$$

and the vanishing of the canonical momentum π^o conjugate to A_o. The non-Abelian generalizations of these equations are well known $|4|$.

In the Hamiltonian framework, the equations $Q_\alpha = 0$ become first class constraints.Quantization of the system then often becomes highly non-trivial in their presence.

2.2 The Lagrangian

The configuration space is usually identified with $F(\bar{D},R)$, where $\bar{D}$ is a fixed time slice of D. It is clear however that for precision we should write $\bar{D}_t$ for the time slice at time t.The customary hypothesis is that $\bar{D}_t$ for different t are diffeomorphic and that there is a natural identification of points of $\bar{D}_t$ for different times.Under these circumstances (which we assume), we are justified in writing $\bar{D}$.

As an example, consider a field theory in Minkowski space M^4. Slices at different times t give different three dimensional subspaces R_t^3.Without further considerations, there is no natural identification of points of these spaces, that is, there is as yet no obvious meaning to the identity of spatial points for observations at different times.What is done in practice is as follows: On M^4, there is an action of the time translation group $\{U_\tau \mid -\infty < \tau < \infty \}$ .The latter maps R_t^3 to $R_{t+\tau}^3$ in a smooth, invertible way.We then identify all points in R_t^3 and $R_{t+\tau}^3$ which are carried into each other by time translation U_τ. In the conventional coordinates $(\vec{x},t)$,

$$U_\tau \ (\vec{x},t) = (\vec{x},\ t + \tau) \tag{2.12}$$

and we think of $\vec{x}$ as referring to the <u>same</u> three dimensional point for all times.

A field f $\epsilon F(D,R)$ restricted to a given time t is a function on $\bar{D}_t$.Since we have an identification of points of $\bar{D}_t$ for different t, the field f can be regarded as a one dimensional family of functions

$f_t \varepsilon \ F(\bar{D},R)$ parametrized by time.We have thus established a correspondence

$$F(D,R) \rightarrow F(R^1, \ F(\bar{D},R)) \tag{2.13}$$

between functions appropiate to the action principle and curves in
the configuration space $F(\bar{D},R)$.

The Lagrangian is a function (or a functional) of "coordinates
and velocities".That is, it is a function of a point $\alpha \ \varepsilon \ F(\bar{D},R)$ on
the configuration space and of the tangent $\dot{\alpha}$ to this space at this
point.This new space (a point of which is a point and a tangent at
the point of the configuration space) is the tangent bundle $T \ F(\bar{D},R)$
on the configuration space.

When the action is reconsructed from the Lagrangian by the formula

$$S \ = \ \smallint dt \ L(\alpha(t), \ \dot{\alpha}(t)) \tag{2.14}$$

we are integrating L along curves in the tangent bundle.This curve
is not arbitrary since we require that $\dot{\alpha}(t) = d\alpha(t)/dt$.Such a curve
in the tangent bundle is the "lift of a curve" from the configuration space.(It is defined by a "second order" vector field in the tangent bundle).With this restriction on curves, a curve on the tangent
bundle is uniquely determined by a curve $\alpha_t \ \varepsilon \ F(\bar{D},R)$. Since such a
curve in turn defines a function in $F(D,R)$, we recover the original
interpretation of the action as a functional on $F(D,R)$.

We need to investigate the action of the gauge group on the
tangent bundle. It turns out that in its action on the tangent bundle, the gauge group, in its simplest version, is associated to the
global group

$$G \ \text{(S)} \ \underline{G} = \{(h,1) \mid h\varepsilon \ G, \ 1\varepsilon \ \underline{G} \} \tag{2.15}$$

where G is the associated global group, and $\underline{G}$ is its Lie algebra and the group multilpication is

$$(h^\sim,l^\sim)(h,l) = (h^\sim h, \ l^\sim + adh^\sim l) \qquad (2.16)$$

Here ad is the adjoint representation of G on $\underline{G}$. In the notation common in physics literature

$$adh^\sim l = h^\sim l \ h^{\sim-1} \qquad (2.17)$$

Thus $G \circledS \underline{G}$ is the semi-direct product of G with $\underline{G}$. This result has been discussed before by Sudarshan and Mukunda [2] .

We denote the gauge group associated to G at a given time by $\hat{G}$. They are functions $F(\bar{D},G) = \{h\}$ with group multiplication defined by

$$(hh^\sim)(\bar{d}) = h(\bar{d})h^\sim(\bar{d}) \ , \ \bar{d}\epsilon\bar{D} \qquad (2.18)$$

The Lie algebra $\underline{G}$ is a group under addition and the associated gauge group at a given time is denoted by $\hat{\underline{G}}$. Finally the gauge group associated to G $\circledS$ $\underline{G}$ at a given time is denoted by $\hat{G} \circledS \hat{\underline{G}}$.

These remarks can be established by examining the way the action of the gauge group "projects down" to an action on coordinates and velocities. A function f $\epsilon F(D,R)$ is transformed to hf. Thus the curve α_t $\epsilon F(\bar{D},R)$ is transformed into $(h\alpha)_t$ where h is time dependent. Thus a point of the tangent bundle is transformed according to

$$(\alpha, \ \dot{\alpha}) \rightarrow (h\alpha, \ d(h\alpha)/dt) = (h\alpha, \ h\dot{\alpha} + l(h\alpha)) \qquad (2.19)$$

where $h \ \epsilon \ \hat{G}$, $l = \dot{h}h^{-1} \ \epsilon \ \hat{\underline{G}}$. In (2.19), the time dependence of h and l have disappeared since we are examining the action at a point of T $F(\bar{D},R)$. In writing (2.19), we have also assumed that the action of the gauge group is local in time, that is

$$(h\alpha)_t = h(t)\alpha(t) \qquad (2.20)$$

If $(h\alpha)_t$ depends on h(t) as well as (say) its derivatives $d^k h(t)/dt^k$, (2.19) will have to be modified. For Yang-Mills theories, this actually happens. (See below). We prefer to illustrate the idea without this complication.With this assumption we can write

$$(h,1) \epsilon \hat{G} \circledS \hat{G} \quad , \quad (h,1)(\alpha,\dot{\alpha}) = (h\alpha, h\dot{\alpha} + 1(h\alpha)) \qquad (2.21)$$

The group multiplication (2.16) follows from

$$(h',1')(h\alpha \quad , \quad h\dot{\alpha} + 1(h\alpha)) = (h'h\alpha, h'h\dot{\alpha} + (h'1h'^{-1})(h'h'\alpha) +$$

$$1'(h'h\alpha)) = (h'h\alpha \quad , \quad h'h\dot{\alpha} + (1' + adh'1)(h'h\alpha)) =$$

$$(h'h, 1' + adh'1)(\alpha,\dot{\alpha}) \qquad (2.22)$$

The preceeding considerations are easily illustrated by Yang-Mills theory where the vector potential A_μ has values in the Lie algebra $\underline{G}$ of the gauge group G and transforms as follows:

$$A_\mu \rightarrow hA_\mu h^{-1} + h\partial_\mu h^{-1} \qquad (2.23)$$

Thus at a fixed time

$$(h,1)A_i = hA_i h^{-1} \qquad (2.24)$$

$$(h,1)A_o = hA_o h^{-1} - 1 \qquad (2.25)$$

where

$$1 = \dot{h}h^{-1} \qquad (2.26)$$

The group multiplication law (2.16) follows by considering the application of $(h',1')$ to the left hand side of (2.24) and (2.25).

The transformation (2.25) on the configuration space variable A_o is not local in time since (2.26) involves dh/dt. Nonetheless, the group multiplication (2.16) is unaffected.

The space on which the group is supposed to act however is not the space of A_μ , but of $(A_\mu, \dot{A}_\mu)$.If we consider the subspace $(A_i, \dot{A}_i)$, since (2.24) does not involve $\dot{h}$, we find the group $\hat{G} \circledS \hat{G}$ .However, the argument has to be modified if $\dot{A}_o$ is considered since its transformation involves $\dot{1}$.An element of the gauge group is now a triple $(h,1,\dot{1})$with the action

$$(h,1,\dot{1})(A_o,\dot{A}_o) = (hA_o h^{-1}-1, h\dot{A}_o h^{-1}+ [1,hA_o h^{-1}] - \dot{1}) \qquad (2.27)$$

and the multiplication law

$$(h_1,l_1,\dot{i}_1)(h_2,l_2,\dot{i}_2) = (h_1h_2,l_1 + h_1l_2h_1^{-1},$$

$$\dot{i}_1 + [l_1,h_1h_2h_1^{-1}] + h_1\dot{i}_2h_1^{-1}) \tag{2.28}$$

The action of $(h,1,\dot{i})$ on $(A_i, \dot{A}_i)$ is obtained from taking the derivative of (2.24).In this action, $\dot{i}$ is passive.

The general gauge group G_L at the Lagrangian level can thus in general involve 1, $\dot{i}$, $\ddot{i}$, ...

The subgroup of <u>constant</u> functions from $\bar{D}$ to G is what is called the global symmetry group.Since it is isomorphic to G, we can denote it by the same symbol G. It is a subgroup of G if there are no boundary conditions on funtions in G, that is if all constant functions are allowed in $\hat{G}$. Thus, with free boundary conditions, we can conclude the following: Since observables are invariant under G, they are invariant under the global group G.That is, all observables are globally neutral.

2.3 The Hamiltonian

The Hamiltonian framework provides an algebraic formulation of the classical theory in terms of Poisson brackets (PB´s).It is an essential step in the quantization of the classical theory [5] .

In this section, we qualititatively describe the preliminaries to Dirac's procedure for setting up the canonical formalism in the presence of constraints. Concrete examples will be worked out in the subsequent chapters.

In the canonical formalism, we start with defining a "cotangent bundle" $T^*F(\bar{D},R)$ on the configuration space $F(\bar{D},R)$ and PB´s between functions on this bundle.This construction is carried out whether or not there are constraints present in the theory. A point in this bundle is labelled by (α,p) where $\alpha \epsilon F(\bar{D},R)$ is a point of the configuration space and p is the conjugate momentum variable. It is also a function on $\bar{D}$.The PB´s involving α and p are conventional.

If we are given a Lagrangian L, then it defines a map $T \ F(\bar{D},R)$
$\to \ T^* F(\bar{D},R)$ by the formula

$$(\alpha, \dot{\alpha}) \ \to \ (\alpha, \frac{\delta L}{\delta \dot{\alpha}}) \qquad\qquad (2.29)$$

The Lagrangian is nonsingular if this map is one-to-one onto
$T^* F(\bar{D},R)$. In that case, when α and $\dot{\alpha}$ range over the allowed values,
all of $T^* F(\bar{D},R)$ is recovered and every point of $T^* F(\bar{D},R)$ uniquely
specifies a state of the system.

It is then an elementary result that the time evolution on
$T^* F(\bar{D},R)$ can be generated by the formula

$$\dot{x} = \{x, H\} \qquad , \ x \ \epsilon T^* F(\bar{D},R) \qquad\qquad (2.30)$$

where H is the Hamiltonian for the system under consideration and
is constructed by the Legendre transform from L and $\{\cdot,\cdot\}$ is the
Poisson bracket.

As we remarked earlier, in gauge theories, the image of the map

$$(\alpha, \dot{\alpha}) \ \to \ (\alpha, \frac{\delta L}{\delta \dot{\alpha}}) \qquad\qquad (2.29)$$

is not all of $T^* F(\bar{D},R)$, but only a submanifold M of this space. That
is, there are functions ϕ_n (n=1,2,...) on $T^* F(\bar{D},R)$ such that ϕ_n is
zero on M:

$$\phi_n (\alpha, \frac{\delta L}{\delta \dot{\alpha}}) \equiv 0 \qquad\qquad (2.31)$$

We note that not all functions on $T^* F(\bar{D},R)$ need to have zero PB's with ϕ_n
on M, that is, $\{f, \phi_n\}$ need not vanish on M for all functions f. For
instance, in electrodynamics the electric field $\vec{E}$ is conjugate to the
potential $\vec{A}$ and Gauss' law

$$\nabla \cdot \vec{E} + j_0 = \phi_1 \qquad\qquad (2.32)$$

is a (secondary) constraint. But obviously,

$$\{A_i (\vec{x}), \phi_1 (\vec{y})\} \ \neq \ 0 \qquad\qquad (2.33)$$

on M. (Note that we must <u>first</u> evaluate the PB´s and <u>then</u> substitute $\phi_n = 0$).Due to the existence of such f, we cannot set $\phi_n = 0$ before evaluating PB´s.Thus we cannot eliminate redundant degrees of freedom using (2.31) without trouble from the Poisson bracket algebra.

A systematic method to treat the constraints is due to Dirac. References 2 contain a detailed exposition of the method. In chapter 8 we will have occasion to illustrate the method in specific examples.

3. Nonrelativistic Particles with Spin

A classical nonrelativistic particle with spin is an example of an elementary system where the utility of the fibre bundle formalism can be illustrated. The Hamiltonian description of such systems is well known [see e.g. Ref.2] and is recalled below. The construction of a Lagrangian description however is not quite straightforward. One such construction involves the use of nontrivial fibre bundles. Below we will only discuss particles with zero electric charge. In chapter 5 we return to a (relativistic) desciption of charged particles.

3.1 The Hamiltonian Description

Let $\vec{x} = (x_1, x_2, x_3)$, $\vec{p} = (p_1, p_2, p_3)$ and $\vec{S} = (S_1, S_2, S_3)$ denote the coordinate, the momentum and the spin of the particle. Here we want to describe a particle with a definite spin. We therefore impose the constraint

$$S^2 \equiv S_i S_i = \lambda^2 \qquad (3.1)$$

where λ is a constant.

The Poisson brackets are

$$\{x_i, x_j\} = \{p_i, p_j\} = 0 \qquad (3.2)$$

$$\{x_i, p_j\} = \delta_{ij} \qquad (3.3)$$

$$\{S_i, S_j\} = \epsilon_{ijk} S_k \qquad (3.4)$$

If the particle is free, the Hamiltonian of the system is

$$H_o = \frac{\vec{p}^2}{2m} \qquad (3.5)$$

where m is the mass of the particle. If there is an external magnetic field $\vec{B} = (B_1, B_2, B_3)$ present and the particle has a magnetic moment μ, the Hamiltonian has the following form:

$$H = H_o + \mu \vec{S} \cdot \vec{B} \qquad (3.6)$$

The equations of motion for the free particle and the inter-acting particle are, respectively

$$\dot{x}_i = p_i/m \qquad\qquad (3.7)$$

$$\dot{p}_i = 0 \qquad\qquad (3.8)$$

$$\dot{s}_i = 0 \qquad\qquad (3.9)$$

and

$$\dot{x}_i = p_i/m \qquad\qquad (3.10)$$

$$\dot{p}_i = - \mu S_j \partial_i B_j \qquad\qquad (3.11)$$

$$\dot{s}_i = \mu \varepsilon_{ijk} B_j S_k \qquad\qquad (3.12)$$

3.2 The Lagrangian Description

If we know the Hamiltonian description, it is often possible to find the Lagrangian of the the system by a Legendre transformation. We can perform the Legendre transformation provided we can find coordinates for the phase space which are canonical. By this we mean the following. Let Q denote the configuration space of the system under consideration. The phase space T^*Q, in our case, is eight-dimensional. A canonical system of coordinates for this space is by definition of the form

$$T^*Q \equiv \left\{ (Q_1, \; Q_2, \; Q_3, \; Q_4, P_1, \; P_2, \; P_3, \; P_4) \right\} \qquad\qquad (3.13)$$

where

$$\{Q_i, \; Q_j\} = \{P_i, \; P_j\} = 0 \qquad\qquad (3.14)$$

and

$$\{Q_i, \; P_j\} = \delta_{ij} \qquad\qquad (3.15)$$

For our system we can, of course, set

$$Q_i = x_i \ , \quad P_i = p_i \ , \quad i \leq 3 \tag{3.16}$$

It remains to find Q_4 and P_4. They will depend on S_α and perhaps on $\vec{x}$ and $\vec{p}$. One can show, however, that there exists no such co-ordinates Q_4 and P_4 which are smooth functions of S. From the constraint (3.1) S spans a 2-dimensional sphere. It is well known that a 2-dimensional sphere cannot be globally coordinatized by a set of coordinates (Q_4, P_4) [6]. Any choice of Q_4 and P_4 will therefore show a singularity for at least one value of S. This singularity is the analogue of the Dirac string [7,8] in the theory of magnetic monopoles. We defer to section 3.3 for a proof of this result.

Thus we cannot find a global Lagrangian by a Legendre trans-formation when we have a constraint like (3.1). (For a local Lagrangian description, see Ref. [9]). Although it is not possible to find a global Lagrangian by a Legendre transformation, the system above does admit a global Lagrangian description. We shall now construct it and point out some of its novel features. The canonical formalism of this Lagrangian is the one discussed above. We will discuss the derivation of this formalism in chapter 8.

Let $\Gamma = \{s\}$ denote the ususal spin 1/2 representation of the rotation group [10]. Thus we have

$$s^\dagger s = 1 \ ; \quad \det s = 1 \tag{3.17}$$

The configuration space Q for the Lagrangian will be the product space $R^3 \times \Gamma$. The points of R^3 as usual correspond to the position coordinates of the particle while a point $s \epsilon \Gamma$ is related to the spin degrees of freedom S_i through

$$S_i \sigma_i = \lambda s \sigma_3 s^{-1} \tag{3.18}$$

Here σ_i, $i = 1,2,3$, are the three Pauli matrices. As a consequence of this definition the constraint (3.1) is fulfilled as an identi-ty.

The Lagrangian of the free spinning particle is

$$L_0 = \frac{1}{2} m\dot{\vec{x}}^2 + \lambda i \mathrm{Tr}(\sigma_3 s^{-1}\dot{s}) \qquad (3.19)$$

where the dot indicates differentiation with respect to time.

We now verify that L_0 gives the correct equations of motion. Variation of the coordinate $\vec{x}$ leads in a known way to

$$m\ddot{x}_i = 0 \ , \ i = 1,2,3 \qquad (3.20)$$

Variation of s can be performed as follows. The most general variation of s can be written in the form

$$\delta s = i \ \varepsilon \cdot \sigma s \qquad (3.21)$$

$$\varepsilon \cdot \sigma = \varepsilon_i \sigma_i \qquad (3.22)$$

This is so because $i\varepsilon \cdot \sigma$ is a generic element of the Lie algebra of Γ and the general variation of s is induced by such an element. Equations (3.17) and (3.21) imply

$$\delta s^{-1} = -is^{-1}\varepsilon \cdot \sigma \qquad (3.23)$$

Hence for the variation (3.21),

$$\delta L_0 = - \lambda \ \mathrm{Tr}(\sigma_3 s^{-1}\dot{\varepsilon} \cdot \sigma \ s) = - 2S_i \dot{\varepsilon}_i \qquad (3.24)$$

After a trivial partial integration, this yields the required equation of motion

$$\dot{S}_i = 0 \qquad (3.25)$$

If the particle has a magnetic moment μ, the Lagrangian in the presence of an external magnetic field $\vec{B}$ is

$$L = L_0 - \frac{\mu}{2} \mathrm{Tr}S\mathcal{B} \equiv L_0 - \mu S_i B_i \qquad (3.26)$$

where

$$S \equiv S_i \sigma_i = \lambda s \sigma_3 s^{-1} \qquad (3.27)$$

and

$$B = B_i \sigma_i \tag{3.28}$$

In (3.26), during variations, we should regard S_i as a function of s. Now the variation of x gives

$$\delta L = m \, \dot{x}_i \, \delta \dot{x}_i - \mu S_j \partial_i B_j \delta x_i \tag{3.29}$$

where

$$\partial_i B_j \equiv \frac{\partial B_j}{\partial x_i} \tag{3.30}$$

Hence

$$m \ddot{x}_i = - \mu S_j \partial_i B_j \tag{3.31}$$

The variation (3.21) of s gives in this case

$$\delta L = - \text{Tr } S \sigma \cdot \dot{\varepsilon} - \frac{i\mu}{2} \text{Tr} \left[S, B \right] \varepsilon \cdot \sigma \tag{3.32}$$

where we have used the cyclic property of the trace operation i.e.

$$\text{Tr } A \left[B, C \right] = \text{Tr } B \left[C, A \right] \tag{3.33}$$

Thus

$$\dot{S}_i = \mu \varepsilon_{ijk} B_j S_k \tag{3.34}$$

Equations (3.31) and (3.34) are the same as those given by the Hamiltonian discussed above.

3.3 Gauge Properties of L_0 and L.

The Lagrangians L_A, A=0,1 where $L_1 \equiv L$, exhibit gauge invariance under a gauge group G which we now discuss in some detail.

Let $U(1) = \{\exp(i\sigma_3 \theta/2)\}$ and consider the transformation

$$s \rightarrow s \exp(i\sigma_3 \theta/2) \tag{3.35}$$

where θ in general is time dependent. Under this transformation, L_A changes only by the time derivative of a function, i.e.,

$$L_A \rightarrow L_A + \lambda \dot{\theta} \tag{3.36}$$

We distinguish this invariance property of a Lagrangian function from the conventional one where the last term in (3.36) is
absent by saying that L_A is weakly invariant under the gauge transformation (3.35). This weak invariance of L_A clearly suggests that
the equations of motion involve only variables invariant under the
gauge transformation (3.35). For dynamical variables, "invariance"
under the transformation (3.35) of course has the conventional
meaning. We may note here that the equations of motion (3.31) and
(3.34) in fact only contain the gauge invariant dynamical variables
x_i and S_i.

Since L_A changes under the gauge transformation (3.35), it
is not possible to write it as it stands in terms of gauge invariant quantities only. We can instead attempt to eliminate the additional gauge degree of freedom in L_A by fixing the gauge. This
means the following: We can show [11] that any gauge invariant
quantity is a function of S_i (and, of course, of x_i). Gauge fixing
means that for each $\vec{S} = (S_1, S_2, S_3)$, we try to find the existence
of an element $s(\vec{S}) \varepsilon$ SU(2) such that

$$S_i \sigma_i = s(\vec{S}) \sigma_3 s(\vec{S})^{-1} \qquad (3.37)$$

If such an $s(\vec{S})$ existed, we could substitute $s(\vec{S})$ for s in
the Lagrangian L_A and therby eliminate the gauge degree of freedom.
We can show, however, that there exists no such choice of $s(\vec{S})$
which is continuous for all $\vec{S}$. The reason for this is as follows.
The vectors $\vec{S}$ which satisfy the normalization condition $S_i S_i = \lambda^2$
span the two sphere S^2. The existence of a smooth $s(\vec{S})$ with the
property (3.37) means that

$$SU(2) = S^2 \times U(1) , \qquad (3.38)$$

since any point in Γ could be written in the form $s(\vec{S}) \exp(i\sigma_3 \theta/2)$.
But SU(2) is simply connected while U(1) on the right hand side of

(3.38) is infinitely connected, and so the right hand side of (3.38) is infinitely connected. (Recall that U(1) is topological-ly identical to the circle S^1). Hence (3.38) and a smooth $s(\vec{s})$ do not exist.

Thus we have the remarkable situation that a Lagrangian for a nonrelativistic spinning system exists only if the space of coordinates and spin variables is nontrivially enlarged to include a U(1) gauge degree of freedom (at least in our approach).

It is often stated in the literature that U(1) gauge invariance implies electromagnetism. But the U(1) gauge invaraiance of the Lagrangian L_A seems to have little to do with electromagnetism. In the sections which follow, we will encounter other weakly gauge invariant Lagrangians in contexts which seem equally remote from Abelian or non-Abelian gauge fields. Thus the assertions in the literature seem to require qualifications.

3.4 Principal Fibre Bundles

The Lagrangians L_A are associated with what in differential geometry is called a principal fibre bundle structure. We now discuss this bundle structure.

As we have seen above, the configuration space appropriate to the Lagrangian L_A is the group space SU(2) = {s}. (More accurately, it is $R^3 \times SU(2)$. But R^3 being, in this case, the set of positions of the particle under consideration, is not relevant in the present context and will be simply ignored.) On the space SU(2), there is the action of the group U(1), i.e. there is an action

$$s \to s \exp(i\sigma_3\theta/2) \qquad\qquad (3.39)$$

Under this action, L_A is weakly invaraiant for time dependent θ's. If we now define the projection map Π by

$$\Pi: SU(2) \to S^2 \qquad\qquad (3.40)$$

where

$$s \to \lambda s \sigma_3 s^{-1} \equiv s_i \sigma_i \tag{3.41}$$

weak invariance of L_A implies that the equations of motion only depend on the base manifold $s^2 = \{\vec{s}\}$.

Thus we have the following mathematical structure:

1) A manifold SU(2) which topologically is the same as the three sphere s^3,

2) the action of a structure group U(1) on the manifold SU(2),

3) the projection map Π from SU(2) to the base manifold s^2. Further,

4) the U(1) action is free, that is, sg=s for g εU(1) implies that g equals the identity element e of the structure group U(1).

Note that the projection Π maps all the right cosets

$$s \, U(1) \equiv s \cdot \{\exp(i \sigma_3 \theta / 2)\} \tag{3.42}$$

to a single point on the base space s^2. This right coset is just the orbit of s under the action of the U(1) group . It is also easy to check that distinct orbits have distinct images on s^2 and that the mapping is onto s^2. That is, the space SU(2)/U(1) of right cosets can be identified with the base space s^2. Thus, if we define an equivalence relation EQ by the statement

$$s_1 \, EQ \, s_2 \text{ if } s_1 g = s_2 \text{ for some } g \, \varepsilon \, U(1) \tag{3.43}$$

then Π is just the map from SU(2) to the space of equivalence classes generated by the relation EQ i.e. SU(2)/U(1).

The preceding features define a principal fibre bundle with the bundle space s^3 ($\equiv$SU(2)) as a manifold, structure group U(1) and base space s^2. It is a well know structure in mathematics –

- the Hopf fibration of the two sphere s^2 [12].

· We now give the general definition of a principle fibre bundle
(For details, see for example, Daniel and Viallet, Ref.1
and reference 13.) It consists of a bundle space E, a structure
group G, a base space B and a projection map Π from E onto B. The
group G = {g} has an action on the bundle space E:

$$E \ni p \rightarrow pg \; \varepsilon \; E \tag{3.44}$$

This action is required to be <u>free</u> i.e.

$$pg = p, \text{ for any } p, \text{ implies that } g \text{ equals the identity } e \text{ of } G \tag{3.45}$$

The projection Π is just the identification of points related
by the G-action. Thus

$$\Pi(p) = \Pi(pg) \tag{3.46}$$

while

$$\Pi(p) = \Pi(q) \tag{3.47}$$

implies that

$$q = pg \tag{3.48}$$

for some $g \; \varepsilon \; G$. We can think of B as the space of G-orbits in E.

A global section is a map

$$\phi: B \rightarrow E \tag{3.49}$$

such that

$$\Pi \circ \phi = \text{identity map on B} \tag{3.50}$$

Thus for $b \; \varepsilon \; B$, $\phi(b)$ is in E and

$$\Pi(\phi(b)) = b \; , \text{ for all } b \text{ in } B \tag{3.51}$$

A local section is defined analogously by restricting the domain
of the defintion of the map B → E to an open set in B. For suit-

able open sets in B, a local section always exists. In fact, there is always a covering $\{V_\alpha\}$ of B by open sets V_α, where $U\ V_\alpha = B$, such that each V_α admits a (local) section.

The principal fibre bundle E is said to be trivial if E = BxG. A principal fibre bundle is trivial if and only if it admits a global section. Note that a point p in a trivial bundle is of the form p=(b,g) where pεB and gεG while the group acts on E as follows:

$$(b,g) \rightarrow (b,gg') , g'\varepsilon G \qquad (3.52)$$

Thus the projection map is just

$$\Pi((b,g)) = b \qquad (3.53)$$

3.5 Gauge fixing

In the conventional treatment of gauge theories (see e.g. Ref. 4), there is a procedure called gauge fixing. It may be explained in the following way. Suppose the configuration space of the Lagrangian under consideration is $\{\xi\}$. Here ξ can be a (multi-component, as well as a space-dependent field. In the latter case, the considerations which follow are only formal. Suppose the gauge group is described by the set $\{g\}$, a time (and possibly space) dependent group, and has the action

$$\xi \rightarrow \xi g \qquad (3.54)$$

on $\{\xi\}$. Fixing the gauge consists of picking exactly one point from each orbit $\xi\{g\}$. This is accomplished by imposing a condition of the form

$$\chi(\xi) = 0 \qquad (3.55)$$

on ξ. $\Big[$ Here χ of course can be multicomponent, $\chi = (\chi_1, \chi_2, \ldots, \chi_n)$. Thus (3.55) can actually be many conditions. $\Big]$ This equation defines a surface M. From the previous remarks, it is clear that the surface M must be such that each orbit cuts M once and exactly once.

If the action (3.54) is free, the previous discussion shows that M is a global section in a principal fibre bundle. In this case, M exists if and only if $\{\xi\}$ is a trivial bundle. Global gauge fixing is possible only in such a case.

In general, the action of the gauge group G on $\{\xi\}$ can be quite involved. Thus a) the action of G may not be free. Then the orbit ξG is not diffeomorphic to G since some elements of G leave ξ fixed, that is, some degrees of freedom of G disappear in the map

$$g \rightarrow \xi g \qquad\qquad (3.56)$$

b) The little group (the stability group, the isotropy group) G_ξ of ξ is the set

$$G_\xi = \{g\epsilon G \mid g = \xi g \} \qquad\qquad (3.57)$$

It may happen that two distinct points ξ and ξ^- have little groups G_ξ and G_{ξ^-} which are not conjugate in G. That is, there exists no element $\bar{g}\epsilon G$, such that

$$\bar{g}G_\xi\bar{g}^{-1} \equiv \{\bar{g}g\bar{g}^{-1} \mid g\epsilon G_\xi\} = G_{\xi^-} \qquad\qquad (3.58)$$

In fact, G_ξ and G_{ξ^-} may not even be isomporphic. An example is the action of the connected Lorentz group $L_+^\uparrow$ on the Minkowski space M^4. For instance, if $x\epsilon M^4$ is timelike, the little group is SO(3), while if x is spacelike the little group is the connected $2 + 1$ Lorentz group. In case b), the different orbits are not diffeomorphic.

If the orbits ξG for different ξ are diffeomorphic, we have a _fibration_ of the space $\{\xi\}$ by the group G. If there are topologically distinct orbits, we have a _singular fibration_ of the space $\{\xi\}$ by the group G.

In Yang-Mills field theory , there are some results which show the non-existence of a global gauge condition [14,15], that

is, of a global surface M with the properties discussed above.
These results are sometimes proved either when the Euclidean
space-time is compactified to the four sphere S^4 or its time
slices are compactified to three spheres S^3.   The physical
meaning of such a compactification is obscure to us [1].

It may be noted that in principle, it is unneccessary to
fix a gauge. The orbits of G in $\{\xi\}$ are well defined. We can
work in the space of these orbits. That is, G defines an equivalence
relation EQ on $\{\xi\}$, ξ and ξ' being equivalent if they are connected by the G-action i.e.

$$\xi \text{ EQ } \xi' \Longleftrightarrow \xi' = \xi g \quad \text{for some } g \varepsilon G \qquad (3.59)$$

The space of orbits is just the space $\{\xi\}$ with G-equivalent
points identified i.e. $\{\xi\}/EQ$. Thus for our spinning system discussed above, it is unnecessary to fix a gauge. In fact, a global
gauge does not exist since the bundle is nontrivial. The space
$\{\xi\}$ in this case is the three sphere S^3, the group G is U(1) and
the space of orbits S^3/EQ is S^2. This example also shows that
even if a global gauge does not exist, the space of orbits, or
the space of gauge invariant variables, still can be well defined.

However, the sort of systems (like the spinning particle)
we discuss in the present work are rather exceptional. Here we
can readily identify the space of gauge invariant variables in
a concrete way. In field theoretical problems, this turns out
to be a difficult problem. The practice in these problems is to
fix the gauge by some convenient procedure. We have seen that a
global gauge fixing is not always possible. Such a circumstance
causes difficulties during quantization.

Recently a perturbation theory for gauge fields without
gauge fixing has been developed [16] based on the Langevin equation of non-equilibrium statistical mechanics. We will not, however, enter into its discussion here.

4. Magnetic Monopoles

In this chapter, we discuss the classical formalism for the description of a nonrelativistic (NR) charged particle in the field of a magnetic monopole [7,8,17]. This system as well illustrates the utility of the fibre bundle formalism in an elementary context. See in this connection Refs. 18-22.

4.1 Equations of Motion

Let $x = (x_1, x_2, x_3)$ denote the relative coordinates and m the reduced mass of the system. We assume that the magnetic field is Coulomb-like. Then the conventional Lorentz force equation reads

$$m\ddot{x}_i = n \frac{1}{r^3} \varepsilon_{ijk} x_j \dot{x}_k \tag{4.1}$$

Here r is the radial coordinate, ε_{ijk} is the Levi-Civita symbol, $4\pi n$ is the product eg of the electric and magnetic charges e and g, and dots denote time differentiation.

The equation (4.1) reveals a remarkable structure when written in terms of radial and angular variables. Let

$$x_i = r\hat{x}_i , \qquad \sum_i (\hat{x}_i)^2 = 1 \tag{4.2}$$

Then (4.1) is equivalent to

$$\ddot{r} = r \sum_i (\dot{\hat{x}}_i)^2 \qquad\qquad (4.3)$$

$$\frac{d}{dt}[\varepsilon_{ijk} x_j (m\dot{x}_k) + n\hat{x}_i] = 0 \qquad\qquad (4.4)$$

The radial equation (4.3) has the same form as for a NR free
particle. But from (4.4), the conserved angular momentum

$$J_i = \varepsilon_{ijk} x_j (m\dot{x}_k) + n\hat{x}_i \qquad\qquad (4.5)$$

has an additional piece $n\hat{x}_i$ as compared to that of the free
particle. It can be interpreted as contributing a helicity

$$\hat{x}_i J_i = n \qquad\qquad (4.6)$$

along the line joining the particle and the monopole.

4.2   The Hamiltonian Formalism

It is much easier to write down a Hamiltonian description of this system than it is to write a Lagrangian description. We describe the former in this section.

Let $B = \{x \in \mathbb{R}^3 | x \neq 0\}$ denote the configuration space.
Note that we have excluded the origin $x = 0$ from B. Thus
the charge and the monopole are forbidden to occupy the same
space-time point. The phase space T*B can be chosen to have

coordinates (x,v) where $v = (v_1, v_2, v_3)$ denotes the relative velocity of the system.

The equation of motion (4.1) is readily verified to be produced by the Hamiltonian

$$H = \frac{1}{2} m \left(\sum_i v_i^2 \right)$$

$$\equiv \frac{1}{2} mv^2 \tag{4.7}$$

provided the Poisson brackets (PB's) are chosen as follows:

$$\{x_i, x_j\} = 0 \tag{4.8}$$

$$\{x_i, v_j\} = \delta_{ij}/m \tag{4.9}$$

$$\{v_i, v_j\} = -\frac{n}{m^2} \varepsilon_{ijk} \frac{x_k}{r^3} \tag{4.10}$$

Note that since the right-hand side of (4.1) is proportional to the magnetic field, the PB (4.10) is conventional for velocities in the presence of a magnetic field.

As was the case for the spinning NR particle, a global Lagrangian can be found if a canonical system of coordinates (Q,P) for T*B can be found. It may again be shown, however, that no such global system of coordinates exists [19]. Thus, it is not possible to construct a global Lagrangian by application of a simple Legendre transformation.

4.3 The Lagrangian Formalism

The global Lagrangian can be constructed by enlarging the configuration space B appropriate to the Hamiltonian to a U(1) bundle E over B. This Lagrangian exhibits a weak gauge invariance under time-dependent U(1) transformations. As a consequence, the equations of motion are defined entirely on B. The structure of the Lagrangian formalism bears a strong resemblance to the one for the nonrelativistic spinning system, although there are important points of difference as well.

Let {s} denote the set of all two-by-two unitary unimodular matrices, i.e. elements of the SU(2) group in the defining representation. The space E is

$$E = \mathbb{R}^1 \times SU(2) \equiv \{(r,s)\} \tag{4.11}$$

Here r is the radial variable with the restriction $r > 0$. So the charge and the monopole are again forbidden to occupy the same space-time point. The relation of s to the relative coordinates x_i is given by

$$\hat{x} = \sigma_i \hat{x}_i$$

$$= s\sigma_3 s^{-1} \tag{4.12}$$

In the Lagrangian below, the basic dynamical variables
are r and s. So, wherever x_i occurs, it is to be regarded as
written in terms of r and s.

The Lagrangian is

$$= \frac{1}{2} m \sum_i \dot{x}_i^2 + ni \, \mathrm{Tr} \, \sigma_3 s^{-1} \dot{s} \qquad (4.13)$$

$$= \frac{1}{2} m\dot{r}^2 + \frac{1}{4} mr^2 \, \mathrm{Tr} \, \dot{\hat{x}}^2 + ni \, \mathrm{Tr} \, \sigma_3 s^{-1} \dot{s} \qquad (4.14)$$

In writing eq. (4.14), the identity $\mathrm{Tr} \, \hat{x}\dot{\hat{x}} = 0$ has been used.
Variation of r in (4.14) leads directly to eq. (4.13). The
most general variation of s is

$$\delta s = i\varepsilon_i \sigma_i s, \qquad \varepsilon_i \text{ real} \qquad (4.15)$$

Hence

$$\delta\hat{x} = i[\varepsilon \cdot \sigma, \hat{x}], \qquad \varepsilon \cdot \sigma = \varepsilon_i \sigma_i \qquad (4.16)$$

$$\delta \, \mathrm{Tr} \, \sigma_3 s^{-1}\dot{s} = i \, \mathrm{Tr} \, \dot{\varepsilon} \cdot \sigma\hat{x} \qquad (4.17)$$

Thus variation of s in (4.14) leads to

$$\mathrm{Tr} \, \varepsilon \cdot \sigma \frac{d}{dt} \left\{ -\frac{1}{2} [\hat{x}, mr^2\dot{\hat{x}}] + n\hat{x} \right\} = 0 \qquad (4.18)$$

where we have used the identity (3.33). The bracketed

expression in (4.18) is a linear combination of Pauli ma-

trices and ε_i is arbitrary.   Therefore,

$$\frac{d}{dt}\left\{-\frac{i}{2}[\hat{x},mr\dot{\hat{x}}] + n\dot{\hat{x}}\right\} = 0 \tag{4.19}$$

which is equivalent to

$$\frac{dJ_i}{dt} = 0 \quad , \tag{4.20}$$

that is, to eq. (4.4).

Thus L leads to both the equations of motion (4.3) and (4.4).

4.4   Gauge Properties of L

The Lagrangian L shows a weak gauge invariance under gauge transformations associated with the U(1) group

$$U(1) = \left\{ g = e^{i\frac{\sigma_3}{2}\theta} \right\} \tag{4.21}$$

This is similar to the weak gauge invariance of the Lagrangians for the spinning systems.   Under the gauge transformation

$$s \longmapsto s\, e^{i\frac{\sigma_3}{2}\theta} \quad , \qquad \theta = \theta(t) , \tag{4.22}$$

we have the weak gauge invariance

$$\mathcal{L} \longmapsto \mathcal{L} - n\dot{\theta} \tag{4.23}$$

As for the spinning system, associated with L, there is the fibre bundle structure

$$U(1) \longrightarrow S^3 \longrightarrow S^2 \qquad (4.24)$$

Again, it is impossible to fix a gauge globally so as to eliminate the U(1) gauge degree of freedom. This is because L is only weakly gauge invariant, and $S^3 \neq S^2 \times U(1)$. Thus there does not exist $s(\hat{x}) \in SU(2)$ which is continuous for all $\hat{x}$ such that

$$\hat{X} = s(\hat{x}) \sigma_3 s(\hat{x})^{-1} \qquad (4.25)$$

It is of course possible to find an $s(\hat{x})$ which fails to be continuous only at one point, say the south pole $[\hat{x} = (0,0,-1)]$. Such an $s(\hat{x})$ is

$$s(\hat{x}) = \frac{1}{2} \left\{ \alpha - \frac{1}{\alpha} [\sigma_3, \hat{x}] \right\} ,$$

$$\alpha = |[2(1 + \hat{x}_3)]^{1/2}| \qquad (4.26)$$

It is easy to verify that $s(\hat{x})$ appearing in eq. (4.26) is a unimodular unitary matrix and fulfills (4.25). Note that $s(\hat{x})$ in (4.26) is not differentiable at the south pole. Substituting (4.26) into the interaction term appearing in eq. (4.13) yields

$$\text{ni Tr } \sigma_3 s(\hat{X})^{-1} \dot{s}(\hat{X}) = n\varepsilon_{3ij}\hat{x}_i \dot{\hat{x}}_j/(1 + \hat{x}_3) \tag{4.27}$$

which is a conventional form of the interaction Lagrangian with a string singularity.

Alternatively, we can cover the two-sphere $S^2 = \{\hat{X}\}$ by two coordinate patches U_1 and U_2 and find group elements $s_i(\hat{X})$ which are defined and continuous in U_i which fulfill (4.25). Substitution of $s_i(\hat{X})$ for s in (4.13) leads to Lagrangians L_i defined on U_i. In the intersection region $U_1 \cap U_2$, in view of (4.25),

$$[s_1(\hat{X})^{-1} s_2(\hat{X})]\sigma_3[s_1(\hat{X})^{-1} s_2(\hat{X})]^{-1} = \sigma_3 \tag{4.28}$$

This means that $s_i(\hat{X})$ differ from each other in the overlapping region by a gauge transformation,

$$s_1(\hat{X}) = s_2(\hat{X}) e^{i\frac{\sigma_3}{2}\theta}$$

for some $\theta = \theta(t)$. Hence L_1 and L_2 differ by the total time derivative of a function in $U_1 \cap U_2$:

$$L_1 = L_2 - n\dot{\theta}$$

Such a (singularity free) formulation which works with two

local Lagrangians is the nonrelativistic analogue of the work of Wu and Yang [21].

5. Relativistic Spinning Particles

In this chapter, we give the Lagrangian description
for relativistic spinning particles, which is formulated on the
Poincare group manifold [23]. It describes a particle which
after quantization is associated with any particular irreducible
representation of the connected Poincare group $P_+^\uparrow$. The
Lagrangian formalism can be generalized [23,24] to include
couplings with both electromagnetism and gravity. We recover
the usual equations of motion for the two systems; i.e. the
Bargmann-Michel-Telegdi [25] equations for electromagnetism
and the Mathisson - Papapetrou [26] equations for gravitation.
The latter equations have been generalized to include coupling to
torsion [27] . Such systems are also recovered from our
formalism.

5.1   The Configuration Space

The Lagrangian is associated with a non-trivial principal
fibre bundle structure which is obtained in the following way.
The bundle space is the connected component of the Lorentz
group. $L_+^\uparrow$ = { Λ }. The structure group is as usual $U(1)$. It
acts on L_+ by right multiplication i.e.

$$\Lambda \quad \to \quad \Lambda g \quad , \quad g\varepsilon\ U(1) \qquad\qquad (5.1)$$

Thus the base space is the space of left cosets $L_+^\uparrow$ / $U(1)$. As
in previous sections we can infer from connectivity arguments
that $L_+^\uparrow$ $\neq$ ($L_+^\uparrow$ / $U(1)$)$\times U(1)$. Thus the bundle is non-trivial.

The configuration space Q for the Lagrangian is the connected Poincare group

$$P_+^\uparrow = \{ (z, \Lambda)\ |\ z=(z^0,z^1,z^2,z^3)\ \varepsilon R^4,\ \Lambda =[\Lambda^a{}_b]\varepsilon\ L_+^\uparrow \}$$

$$\qquad\qquad\qquad\qquad\qquad\qquad\qquad\qquad (5.2)$$

Here z^a is interpreted as the space-time coordinate of the
particle. The interpretation of Λ is as follows. If p_a and S_{ab}

are the momentum and spin components of the particle,

$$p_a = m \Lambda_{a0} \quad , \quad m > 0 \tag{5.3}$$

$$\frac{1}{2} S_{ab} \sigma^{ab} = \lambda \Lambda \sigma_{12} \Lambda^{-1} \equiv -iS \tag{5.4}$$

where

$$(\sigma^{ab})_{cd} = -i(\delta^a_c \delta^b_d - \delta^a_d \delta^b_c) \tag{5.5}$$

and λ is a constant. These equations are valid for a time-like four vector p_a. The cases where the four vector p_a is not time-like will be treated later. Note that by the definitions above,

$$S_{ab} = \lambda(\Lambda_{a1}\Lambda_{b2} - \Lambda_{a2}\Lambda_{b1}) \tag{5.6}$$

and

$$p_0 = m\Lambda_{00} > 0 \quad , \quad p_a p^a = -m^2 \tag{5.7}$$

Therefore we obtain

$$\frac{1}{2} S_{ab} S^{ab} = \lambda^2 \tag{5.8}$$

and

$$p_a S^{ab} = 0 \tag{5.9}$$

Here the Latin indicies are raised and lowered by the Lorentzian metric

$$\eta = (-1,1,1,1) \tag{5.10}$$

5.2 The Lagrangian for a Free Spinning Particle

The Lagrangian for a massive spinning particle is

$$L_p = p_a \dot{z}^a + i\frac{\lambda}{2} \operatorname{Tr}(\sigma_{12}\Lambda^{-1}\dot{\Lambda}) \tag{5.11}$$

where p_a is defined in the equation (5.3) The variables are functions of the parameter τ which parametrize the space-time trajectory of the particle. The dot indicates differentiation with respect to τ. Note that $\int L_p d\tau$ is invariant under reparametrizations, $\tau \to f(\tau)$.

Let us first derive the equations of motion. The variation of z^a is standard and leads to

$$\dot{p}_a = 0 \tag{5.12}$$

The most general variation of Λ is, as usual,

$$\delta \Lambda = i \ \varepsilon \cdot \sigma \Lambda \qquad (5.13)$$

where

$$\varepsilon \cdot \sigma = \varepsilon^{ab} \sigma_{ab} \qquad (5.14)$$

This implies

$$\delta \Lambda^{-1} = -i \Lambda^{-1} \varepsilon \cdot \sigma \qquad (5.15)$$

Hence

$$\delta L_p = -i \text{Tr} \, (k \varepsilon \cdot \sigma) + \frac{i}{2} \text{Tr} \, (S \frac{d}{d\tau} (\varepsilon \cdot \sigma)) \qquad (5.16)$$

where the matrix k is defined by $k_{ab} = z_a p_b$. The traces have a conventional meaning i.e.

$$\text{Tr} \, (k \varepsilon \cdot \sigma) = \sum_{a,b} k^{ab} (\varepsilon \cdot \sigma)_{ba} \qquad (5.17)$$

After a partial integration in the equation (5.16) and use of (5.12), we obtain the equation for the conservation of total angular momentum:

$$\frac{d}{d\tau} M^{ab} = 0 \qquad (5.18)$$

where

$$M^{ab} = z^a p^b - z^b p^a + S^{ab} \qquad (5.19)$$

The proof that L_p actually describes a particle which is associated with an irreducible representation of the connected Poincare group $P^{\uparrow}_+$ follows by showing that mass and spin have definite values. The mass has a definite value due to eq. (5.7). Note also that the sign of p_0 is fixed by (5.3) since $\Lambda \varepsilon \ L^{\uparrow}_+$. Thus L_p does not describe a particle which can have both positive and negative energies. Both signs can be obtained by abandoning the condition that $\Lambda_{00} > 0$.

We can show that the spin has a definite value from (5.8) and (5.9). The latter shows that in the rest frame of the particle, the spin tensor S^{ab} has only space components. The former shows that the magnitude of this spin tensor has a definite numerical value. Thus the spin 3-vector $S_i = \varepsilon_{ijk} S_{jk}$ has a definite value in the particle rest frame. In general, by computing the square of the Pauli-Lubanski vector W_a,

$$W_a = \frac{1}{2} \varepsilon_{abcd} M^{bc} p^d \tag{5.20}$$

where ε_{abcd} is the usual antisymmetric tensor with $\varepsilon_{0123} = 1$, we find

$$W_a W^a = m^2 \lambda^2 \tag{5.21}$$

It is important to realize that the preceding equations imply

$$p_a = m\dot{z}_a / (-\dot{z}^2)^{1/2} \tag{5.22}$$

and

$$\dot{S}_{ab} = 0 \tag{5.23}$$

Thus the conventional relation between momentum and velocity is recovered, and (5.12) becomes the usual equation of motion when written in terms of z_a. The derivation of these results relies on (5.18) which can be rewritten as

$$\dot{z}^a p^b - \dot{z}^b p^a + \dot{S}^{ab} = 0 \quad , \tag{5.24}$$

in view of (5.12). It also relies on the time derivative of (5.9), i.e.

$$p_a \dot{S}^{ab} = 0 \tag{5.25}$$

Multiplication of (5.24) by p_a shows that p_a and $\dot{z}_a$ are, in fact, parallel. The constant of proportionality can be determined by using the normalization condition $p_a p^a = -m^2$ and the condition that $p_0 > 0$. This gives (5.22). Now (5.22) applied to (5.24) yields (5.23).

The canonical quantization of the Lagragian (5.11) will be carried out in section 8.3.

5.3 The Spinning Particle in an Electromagnetic Field

We now discuss the coupling of electromagnetism to spinning particles [24]. In order that our system reduces to the standard formulation in the limit of zero spin, the minimal coupling term $eA_a(z)\dot{z}^a$ must be present in the interaction Lagrangian. Here A_a is

the electromagnetic potential. When the spin is non-zero, an additional coupling to the electromagnetic field of the form $cF_{ab}(z)S^{ab}$
may be present. As we will see below, the constant c is associated with the gyromagnetic ratio of the particle. One possible choice for the electromagnetic interaction Lagrangian is
therefore

$$L_{EI} = eA_a(z)\dot{z}^a + c\sqrt{-\dot{z}^2}\ F_{ab}(z)S^{ab} \qquad (5.26)$$

The second term in (5.26) is the generalization of the interaction term in the Hamiltonian (3.6) for a non-relativistic,
spinning particle. The factor $\sqrt{-\dot{z}^2}$ in the second term of (5.26)
was inserted in order to retain the reparametrization symmetry
$\tau \rightarrow f(\tau)$. As will be shown later, alternatives to (5.26) are
possible.

The equations of motion are obtained by varying Λ and z in
the total action

$$S = \int d\tau\ L_P + \int d\tau\ L_{EI} \qquad (5.27)$$

Variations in Λ now lead to

$$\dot{S}^{ab} + \dot{z}^a p^b - \dot{z}^b p^a = c\sqrt{-\dot{z}^2}\ (\ F^{ac}(z)S_c^{\ b} - F^{bc}(z)S_c^{\ a}\ ) \quad , \qquad (5.28)$$

where M^{ab} is defined in (5.19). Variations of z^a leads to

$$\dot{p}_a = eF_{ab}\dot{z}^b + c\sqrt{-\dot{z}^2}\ S_{cd}\partial_a F^{cd} + \frac{d}{d\tau}(\ \frac{c\dot{z}_a}{\sqrt{-\dot{z}^2}}\ S{\cdot}F(z)\ ) \quad , \qquad (5.29)$$

where we have introduced the notation $S{\cdot}F = S_{ab}F^{ab}$.
Note that we no longer have the usual relationship between
momentum and velocity (5.22). In general, the velocity and
momentum are not even parallel. This follows after substitution of (5.28) and (5.29) into the condition

$$\dot{S}^{ab}P_b + S^{ab}\dot{P}_b = 0 \qquad\qquad (5.30)$$

We find,

$$P_a = -\frac{1}{(p\cdot\dot{z})}\left(m^2\dot{z}_a + c\sqrt{-\dot{z}^2}\left(P_bF^{bc}S_{ca} + S_a{}^bS_{cd}\partial_bF^{cd}\right)\right.$$
$$\left. + cS_{ab}\frac{d}{d\tau}\left(\frac{\dot{z}^b}{\sqrt{-\dot{z}^2}}\, S\cdot F\right) + eF_{bc}\dot{z}^bS^{ab}\right) \qquad (5.31)$$

In order to compare this system of equations with that of Bargmann et al. [25], let us examine the weak homogeneous field limit. Upon substituting (5.31) and (5.28) into (5.29), we find the Lorentz equation of motion

$$m\frac{d}{d\tau}\left(\frac{\dot{z}^a}{\sqrt{-\dot{z}^2}}\right) = eF^{ab}\dot{z}_b \qquad\qquad (5.32)$$

The equation for the spin precession can be given in terms of the Pauli-Lubanski vector (5.20). Substituting (5.28), (5.29) and (5.31) into the equation

$$\dot{W}_a = \frac{1}{2}\,\varepsilon_{abcd}\left(\dot{S}^{bc}P^d + S^{bc}\dot{P}^d\right) \qquad\qquad (5.33)$$

and again keeping terms which are at most linear in the homogenous field we find

$$\dot{W}_a = -2c\sqrt{-\dot{z}^2}\, F_{ab}W^b - \frac{2c + e/m}{\sqrt{-\dot{z}^2}}\left(\dot{z}^cF_{cb}W^b\right)\dot{z}_a \qquad (5.34)$$

Equations (5.32) and (5.34) are the Bargman-Michel-Telegdi equations for a spinning particle, with the identification

$$c = -\frac{eg}{4m} \qquad\qquad (5.35)$$

g being the gyromagnetic ratio.

The field equations for this system are obtained by adding the usual free field action

$$S_E = - \frac{1}{4} \int d^4x \; F_{ab}(x) F^{ab}(x) \tag{5.36}$$

to (5.27). By varying the electromagnetic potentials A_a and integrating by parts, we find

$$\partial_a F^{ab}(x) = - e \int d\tau \delta^4(x-z(\tau)) \dot{z}^b + 2c \int d\tau \partial_a \delta^4(x-z(\tau)) \cdot$$
$$\sqrt{-\dot{z}^2} \; S^{ab} \tag{5.37}$$

where, as above, we use the notation $\partial_a \equiv \partial/\partial x^a$. The second term on the right hand side of (5.37) represents the dipole contribution to the field in the sense of Papapetrou [27] (see also in this context the work by Bailyn and Ragusa [28] and references therein).

As was stated above the interaction Lagrangian (5.26) is not uniquely determined (for a related discussion see Ref. [29]). For instance, we can replace the second term in (5.26) by [30]

$$- \frac{c}{m} P_a \dot{z}^a S_{bc} F^{bc}(z) \tag{5.38}$$

This term preserve all the symmetries of the previous system, yet it gives a different set of equations of motion. In the limit of a weak homogeneous field the two systems can, however, be shown to be equivalent. Note that the term (5.38) can be absorbed in the first term in L_P (5.11), through a "renormalization" of the mass m:

$$m \to M(\alpha) = m + \frac{eg}{4m}\alpha \; ; \; \alpha = F_{ab} S^{ab} \tag{5.39}$$

In fact, if we no longer restrict ourselves to Lagrangians which are linear in $F_{ab}S^{ab}$, we can consider the case where the mass M is an arbitrary function of α (which may be relevant when one is considering particles with an anomalous magnetic moment [31]). In this case the total particle Lagrangian would be

$$L_p + L_{EI} = M(\alpha) \; \Lambda_{a0} \dot{z}^a + i\frac{\lambda}{2} \, Tr\sigma_{12}\Lambda^{-1}\dot{\Lambda} + eA_a\dot{z}^a \qquad (5.40)$$

The resulting equations of motion are

$$S^{ab} + \dot{z}^a p^b - \dot{z}^b p^a = \frac{d}{d\alpha}(\ln M) p_c \dot{z}^c (S^a{}_c F^{cb} - S^b{}_c F^{ca}) \qquad (5.41)$$

and

$$\dot{P}_a = eF_{ab}\dot{z}^b + \frac{d}{d\alpha}(\ln M) p_c \dot{z}^c \partial_a F_{bd} S^{bd} \qquad (5.42)$$

(These equations have also been considered by Dixon [32]). Even though the equations (5.41) and (5.42) correspond to a large class of systems, depending on the choice of $M(\alpha)$, they all lead to the Bargmann-Michel-Telegdi equations in the weak and homogeneous field limit. Here the identification of the particle´s mass and the gyromagnetic ratio g are given by

$$m = M(0) \qquad (5.43)$$

$$g = \frac{4}{e}M(0)\frac{dM(\alpha)}{d\alpha}\Big|_{\alpha=0} \qquad (5.44)$$

which is, of course, consistent with the specific choice (5.35).

5.4 The Spinning Particle in a Gravitational Field.

It is rather straightforward to generalize L_p to include gravitational effects. It is then convenient to regard the gravitational field as a gauge field [33] , i.e. the Poincare group is regarded as a local symmetry group. Let $h = (h_\mu^a)$ be the vierbein fields and A_μ^{ab} the Yang-Mills potentials for the Lorentz group. (Our notation is as follows. A Latin index like a is a tangent space index and a Greek index like μ is a curved space index.) The action now is

$$S = \int d\tau\, L_P + \int d^4x\, L_F \qquad (5.45)$$

where

$$L_P = m\Lambda_{a0} h^a_{\ \mu} \dot{z}^\mu + i\frac{\lambda}{2}\, Tr(\sigma_{12}\Lambda^{-1}D_\tau\Lambda) \qquad (5.46)$$

and

$$L_F = -\frac{1}{16\pi G}\, F^{ab}_{\mu\nu} h^\mu_a h^\nu_b \, det(h^c_\rho) \qquad (5.46)$$

Here the Yang-Mills field strength $F^{ab}_{\mu\nu}$ are given by

$$F_{\mu\nu} = \frac{i}{2} F^{ab}_{\mu\nu}\sigma_{ab} = [D_\mu, D_\nu] \qquad (5.47)$$

where D_μ is the covariant derivative

$$D_\mu = \partial_\mu + \frac{i}{2} A^{ab}_\mu \sigma_{ab} \qquad (5.48)$$

and

$$D_\tau\Lambda = \dot{\Lambda} + \dot{z}^\mu A_\mu \Lambda \qquad (5.49)$$

Futhermore, in (5.46), G is Newton's constant.

The equations of motion are found as follows. If we vary as in (5.13), (5.16) is replaced by

$$\delta L_P = -i\, Tr(J\varepsilon.\sigma) + \frac{i}{2}\, Tr(S\frac{d}{d\tau}(\varepsilon.\sigma) - [\dot{z}^\mu A_\mu, S]\varepsilon.\sigma) \qquad (5.50)$$

where

$$J^{ab} = h^{a\mu} \dot{z}_\mu (m\Lambda^{b0}) \qquad (5.51)$$

We thus find the equation for spin precession [34]

$$J^{ab} - J^{ba} + (D_\tau S)^{ab} = 0 \qquad (5.52)$$

where

$$D_\tau S = \frac{dS}{d\tau} + [\dot{z}_\mu A^\mu, S] \qquad (5.53)$$

Variation of the coordinate z_μ leads to the Mathisson-Papapetrou equation in the presence of torsion [26,27]. We find

$$\delta L_p = - \delta z^\mu (p_\mu - z_\lambda (\partial_\mu h^{a\lambda}) p_a + \delta z^\mu \frac{1}{2} \, \text{Tr}(SA_\mu)$$

$$- \delta z^\mu \frac{1}{2} \, \text{Tr}(S \partial_\mu A_\lambda) \tag{5.54}$$

where $p_\mu = h^a_\mu p_a$. Partial integration in the second term and substitution from the equation (5.52) leads to

$$\dot{p}_\mu - \dot{z}_\lambda (\partial_\mu h^{a\lambda}) p_a + \dot{z}_b A^{ba}_\mu p_a - \frac{1}{2} \, \text{Tr}(SF_{\mu\nu} \dot{z}^\nu) = 0 \tag{5.55}$$

This is the same equation as the Mathisson-Papapetrou equation in the presence of torsion i.e.

$$(D_\tau p)_a - h^\mu_a \dot{z}^\nu ((D_\mu h_\nu)^b - (D_\nu h_\mu)^b) p_b$$

$$- \frac{1}{2} h^\mu_a \dot{z}^\nu \text{Tr}(SF_{\mu\nu}) = 0 \tag{5.56}$$

In the Equation (5.56) we make use of the notation

$$(D_\tau p)_a = \frac{dp_a}{d\tau} + \eta_{ab} \dot{z}^\lambda A^{bc}_\lambda p_c \tag{5.57}$$

and

$$(D_\mu h_\nu)^b = \partial_\mu h^b_\nu + A^{bc}_\mu h_{c\nu} \tag{5.58}$$

For a discussion of the field equations, we refer to Ref. 23.

5.5 General Irreducible Representations of $P^\uparrow_+$.

To find Lagrangian descriptions for other unitary irreducible representations of the Poincare group $P^\uparrow_+$, it is sufficient to alter the definitions of p_a and S_{ab}. For example, in order to describe a massless particle, like the photon or the neutrino, we may replace $m\Lambda_{a0}$ in (5.7) by

$$(\Lambda_{a0} + \Lambda_{a3}) \, \omega \tag{5.59}$$

where ω corresponds to the "frequency" of the massless particle. Equation (5.59) ensures that

$$P_a P^a = 0 \qquad (5.60)$$

For a massless particle the Pauli-Lubanski vector W_a, as given
by the Equation (5.20), obeys the condition that $W_a W^a = 0$, and
since $p_a W^a = 0$, it is easy to show the following identity

$$W_a = \lambda P_a \qquad (5.61)$$

It follows that for $\lambda = 1/2$ and $\lambda = 1$, we get a "neutrino" and
a "photon" of definite helicity. The sign of the helicity can
,of course, be reversed by reversing the sign of λ .

The tachyonic representations are obtained by choosing

$$P_a = \rho \Lambda_{a3} \qquad (5.62)$$

Different values of ρ and λ give different irreducible representations as may be seen from the values of the invariants
$P_a P^a$ and $W_a W^a$:

$$P_a P^a = \rho^2 \qquad (5.63)$$

$$W_a W^a = -\rho^2 \lambda^2 \qquad (5.64)$$

For the irreducible representations with zero four-momentum,
we set

$$P_a = 0 \qquad (5.65)$$

and choose the Lagrangian to be

$$L_p = \frac{i}{2} \, \text{Tr} (K \Lambda^{-1} \dot{\Lambda}) \qquad (5.66)$$

where $K = K_{ab} \sigma^{ab}$ is a __fixed__ element of the Lie algebra. From
the variation (5.13), we find the equation of motion

$$\frac{d}{d\tau} S_{ab} = 0 \qquad (5.67)$$

where

$$\frac{1}{2} S_{ab} \sigma^{ab} = \Lambda K \Lambda^{-1} \qquad (5.68)$$

Thus S is the spin angular momentum. The irreducible representations are characterized by the two invariants [35]

$1/2\ S_{ab}S^{ab}$ and $1/2\varepsilon^{abcd}S_{ab}S_{cd} \equiv S_{ab}{}^{*}S^{ab}$. Since

$$\frac{1}{2}\ S_{ab}S^{ab} = 2K_{ab}K^{ab} \tag{5.69}$$

and

$$\frac{1}{2}\ S_{ab}{}^{*}S^{ab} = 2K_{ab}{}^{*}K^{ab} \tag{5.70}$$

we can classically get any real values for these invariants
by choosing valuer for K_{ab} appropriately. To quantize the system
(see chap.8), we are, as usual, obliged to give them values which
are appropriate for unitary irreducible representations [35].

5.6 Relation Between the Charge-Monopole System and the Mass-less Spinning Particle System

In this section we point out some striking analogies between
the charge-monopole system and the system of a massless particle
of fixed helicity. The similarities of the two systems become
evident when the roles of coordinates and velocities are inter-
changed.

The analogies are as follows:
1) The angular momentum of a charged particle in the field of
monopole contains a helicity n (see chapter 4, the equation (4.6))
along the direction joining the monopole and the charge. The
angular momentum of a massless particle of spin λ contains
helicity λ along the direction of the momentum of the particle.

2) The components of the position vector of the charge-monopole
system commute, but the components of the velocity vector do
not (at least not for finite charge-monopole separation). Thus
the system cannot be localized in velocity space. Furthermore,
there is no globally defined momentum vector, consequently a

globally defined momentum space wave function cannot be defined. For a massless particle, on the other hand, the components of momenta commute i.e.

$$[p_i, p_j] = 0 \quad , \tag{5.71}$$

but the components of position do not

$$[x_i, x_j] = - i\lambda \epsilon_{ijk} p_k / p^3 \tag{5.72}$$

Using equations (5.71) and (5.72), along with the canonical commutation relation (5.75) we can verify that $J_i = \epsilon_{ijk} x_j p_k + \lambda \hat{p}_i$ generates rotations for this system. Equation (5.72) is analogous to the commutation relation for the components of velocity for the charge-monopole system (4.10). It is consistent with the fact that the photon cannot be localized [36] . With the Hamiltonian $H = |\vec{p}|$, we are lead to the equations of motion

$$\ddot{x}_i = 0 \tag{5.73}$$

and

$$\frac{d}{dt}(\epsilon_{ijk} x_j p_k + \lambda \hat{p}_i) = 0 \tag{5.74}$$

if we supplement the commutation relations (5.71) and (5.72) with the canonical commutation relation

$$[x_i, p_j] = i\delta_{ij} \tag{5.75}$$

3) The non-trivial topology of the charge-monopole system depends on the fact that their relative spatial separation cannot become zero. As a consequence, the configuration space has the topology $R^1 \times S^2$. The unusual topological features of the charge-monopole system can be characterized in terms of this bundle. If the relative coordinate is allowed to vanish as well, the configuration space becomes R^3, which does not admit non-trivial U(1) bundles.

In contrast, since for a massless particle its three momentum cannot be transformed to zero by Lorentz transformations, the origin in momentum space should be excluded. The topology of $\{ \vec{p} \}$ is thus $R^1 x S^2$. For a non-zero helicity, its Lagrangian description is faciliated by making use of the $U(1)$ bundle $R^1 x S^3$ over $R^1 x S^2$. In the photon Lagrangian, the entire Lorentz group appears to play the role of the bundle space. Consider, however, the translation group T_2 as generated by

$$\Pi_1 = M_{10} + M_{13} \tag{5.76}$$

and

$$\Pi_2 = M_{20} + M_{23} \tag{5.77}$$

The photon Lagrangian is invariant under the transformations

$$\Lambda \to \Lambda \cdot \exp(i\alpha^a(x)\Pi_a) \tag{5.78}$$

Thus it can be globally written on $L_+/T_2 = R^1 x S^3$ by factoring these gauge degrees of freedom. The Euclidean group generated by σ_3, Π_1 and Π_2 is the familiar stabilty group of the four momentum $(1,0,0,1)$.

4) From the expression for the conserved angular momentum, we see that under a parity transformation, $\lambda \to -\lambda$ for both systems under consideration. Thus a charge-monopole system with a fixed value of e and g [e(g) being the electric (magnetic) charge], or a massless particle of fixed helicity, is incompatible with parity invariance .

5) There are no bound states in the charge-monopole system. For large times, the motion approaches that of two free particles. That is, as $t \to \infty$,

$$\vec{x}(t) \rightarrow \vec{v}t + \vec{x}_0 + O(t^{-2}) \qquad (5.79)$$

where $\vec{x}(t)$ is the trajectory in the relative coordinate, and $\vec{x}_0$ and $\vec{v}$ are constant vectors. It follows that as $t \rightarrow \infty$, the conserved angular momentum $J_i = \varepsilon_{ijk}x_k p_k + \lambda\hat{x}_i$, where $\vec{p} = m\vec{v}$, approaches the value

$$J_i = \varepsilon_{ijk}(\vec{x}_0)_j p_k + \lambda\hat{p}_i \qquad (5.80)$$

which has the same form as that for a massless particle. In Ref. [37], the preceeding limit for the charge-monopole system was discussed in detail. It was shown that the commutation relations of $\vec{x}_0$ and $\vec{p}$ are the same as those in the Equations (5.71), (5.72) and (5.75).

A canonical formalism was, futhermore, developed in Ref. [37] for a free, non-relativistic particle with no internal degrees of freedom. This formalism was unusual in that upon quantization, the angular momentum contained a helicity λ in the direction of the three mommentum $\vec{p}$. If λ is chosen as half integral, the system thus becomes "fermionic". Such a system resembles a massless particle or the large time limit of the charge-monopole system. In Ref. [37] no Lagrangian formulation of the system was given. We may notice here that it is just the non-relativistic analogue of the photon Lagrangian i.e.

$$L = p\hat{p}_k\dot{x}_k - i \operatorname{Tr}(\sigma_3 s^{-1}\dot{s}) \qquad (5.81)$$

Here $s \in SU(2)$, the momentum is $p_k = p\hat{p}_k$ and $\sigma_k p_k = s\sigma_3 s^{-1}$. Thus p is not an independent variable, but is defined in terms of the dynamical group element s.

6. Yang-Mills Particles

The classical description of a charged particle in an electromagnetic field is well known. The motion of the particle is described by the Lorentz force equation, while the dynamics of the field is described by the Maxwell equations. The non-Abelian generalization of these equations is due to Wong [38]. Instead of an electric charge, the corresponding Yang-Mills particle carries a spin-like variable $\vec{I}$ which transforms under the adjoint representation of the internal symmetry group. The Wong equations provide a coarser level of description than a non-Abelian gauge field theory since they treat the sources only as particles. Hence they may be more tractable than a gauge field theory and may also reveal important features of the latter. For such reasons, there is currently a growing interest in the Wong equations.

Below, we first recall the Wong equations. Then the Hamiltonian and Lagrangian descriptions of these equations are discussed. The Lagrangian description in our approach [11,39] requires the use of nontrivial fiber bundles.

6.1 The Wong Equations

The Wong equations are

$$m \frac{d}{d\tau} \left[\frac{\dot{z}_a}{\sqrt{-\dot{z}^2}} \right] = -e \mathcal{F}^{\alpha}_{ab} (z) I_{\alpha} \dot{z}^b \tag{6.1}$$

$$(D^{a})_{\alpha\beta} \mathcal{F}^{\beta}_{ab}(x) = e\int d\tau \delta^{4}[x - z]\dot{z}_{b}I_{\alpha} \tag{6.2}$$

Here, $z^{a} = z^{a}(\tau)$ denotes the particle trajectory in Minkowski space, while $\mathcal{F}^{\alpha}_{ab}$ and D^{a} are the usual Yang-Mills tensor and co-variant derivative, respectively. The range of the indices α and β is equal to the dimension n of the internal symmetry group G. The vector $\vec{I} = \vec{I}(\tau)$ transforms under the adjoint representation of G. From (6.2) and the identity $[D^{a},D^{b}]_{\alpha\beta}\mathcal{F}^{\beta}_{ab} = 0$ one finds the following consistency condition on I:

$$\frac{d}{d\tau} I_{\alpha}(\tau) - e\dot{z}^{a}A^{\rho}_{a}c_{\rho\alpha\beta}I_{\beta} = 0 \tag{6.3}$$

Here, $c_{\rho\alpha\beta}$ are the structure constants.

It is known [11] that the spectrum of the Casimir invariants constructed out of $\vec{I}$ determines the irreducible representations (IRRs) of G which occur in the quantum theoretic Hilbert space. We want to describe a particle which belongs to a definite IRR of G. Thus we impose also the constraint:
Casimir invariants of $\vec{I}$ have definite numerical values.
It is easy to show that this constraint is consistent with the time evolution of $\vec{I}$ (6.3).

It is instructive to verify that the preceding equations reduce to the Lorentz and Maxwell equations when $G = U(1)$. In this case, $\vec{I} = I_{1}$ has one component. Since $c_{\rho\alpha\beta} = 0$,

I_1 is a constant of motion by (6.3) and it can be assigned a definite numerical value, say λ. Identifying $e\lambda$ with the electric charge, (6.1) and (6.2) are seen to reduce to Lorentz and Maxwell equations.

6.2 The Hamiltonian Formalism

The Hamiltonian is a simple generalization of the electrodynamic Hamiltonian. It is

$$H = H_p + H_F \tag{6.4}$$

where

$$H_p = \{[p_i - eA_i^\alpha(z)I_\alpha]^2\}^{\frac{1}{2}} + eA_0^\alpha(z)I_\alpha \tag{6.5}$$

and H_F is the Hamiltonian for the Yang-Mills field [4]. The latter is well known. In writing (6.5), we have identified z_0 with time $\tau \equiv t$. The Poisson brackets (PBs) involving z_i's and p_i's (i = 1,2,3) are conventional. They have also zero PBs with I_α. The PBs involving I_α alone are

$$\{I_\alpha, I_\beta\} = c_{\alpha\beta\gamma}I_\gamma \tag{6.6}$$

It is straightforward to verify that this Hamiltonian

and the PBs lead to the required equations of motion.

6.3 The Lagrangian Formalism

The presence of a spin-like variable $\vec{I}$ whose Casimir invariants are fixed suggests in analogy to previous sections that a Lagrangian can be found on a configuration space E which contains additional gauge degrees of freedom. This is indeed the case. The space E turns out to be $\mathbb{R}^3 \times G$ where $\mathbb{R}^3$ is the usual space of spatial coordinates and G is the internal symmetry group.

We assume as usual that G is a compact connected Lie group with a simple Lie algebra $\underline{G}$. Let $\Gamma = \{s\}$ be a faithful unitary representation of G. The associated Lie algebra $\underline{\Gamma}$ has a basis $T(\rho)$ $(\rho = 1,2,\ldots,n)$ with $T(\rho)^\dagger = T(\rho)$. [More precisely, this is a basis for $i\underline{\Gamma}$.] We choose $T(\rho)$ so that the normalization condition

$$\text{Tr } T(\rho)T(\sigma) = \delta_{\rho\sigma} \tag{6.7}$$

is fulfilled. The commutation relations of $T(\rho)$ are

$$[T(\rho),T(\sigma)] = ic_{\rho\sigma\lambda}T(\lambda) \tag{6.8}$$

The Lagrangian for the particle dynamics is

$$L = -m[-\dot{z}(\tau)^2]^{\frac{1}{2}} - i \text{ Tr } [Ks^{-1}D_\tau s] \tag{6.9}$$

Here $s = s(\tau) \in \Gamma$ represents the novel degrees of freedom in L. The covariant derivative D_τ is defined by

$$D_\tau = \frac{d}{d\tau} - ie\dot{z}^a A_a \text{ ,}$$

$$A_a \equiv A_a^\alpha[z(\tau)]T(\alpha) \tag{6.10}$$

where A_a^α are the Yang-Mills potentials. The matrix K is defined by

$$K = K_\rho T(\rho) \tag{6.11}$$

where K_ρ are real valued <u>constants</u>. Their specific values determine the IRR of G to which the particle belongs.

The Yang-Mills Lagrangian

$$-\frac{1}{4}\int d^3x \; \mathcal{F}_{ab}^\alpha \mathcal{F}^{\alpha,ab} \tag{6.12}$$

can be added to L. We omit it here since the treatment of the Yang-Mills field is standard.

The definition of the internal vector $\vec{I}$ in terms of s and K is

$$I = I_\alpha T(\alpha)$$

$$= sKs^{-1} \tag{6.13}$$

The resemblance of (6.9) and (6.13) to the corresponding equations in the previous sections should be noted.

Let us now derive the equations of motion. The general variation of s is, as usual,

$$\delta s = i\varepsilon \cdot T \ s$$

$$\varepsilon \cdot T = \varepsilon_\alpha T(\alpha) \tag{6.14}$$

For this variation,

$$\delta L = \text{Tr } K \, s^{-1}\{\dot{\varepsilon} \cdot T + ie[\varepsilon \cdot T, \dot{z}^a A_a]\}s \tag{6.15}$$

This becomes after a partial integration

$$- \text{Tr } \varepsilon \cdot T\{D_\tau I\} \tag{6.16}$$

where

$$D_\tau I \equiv \frac{dI}{d\tau} - ie[\dot{z}^a A_a, I] \tag{6.17}$$

Since $D_\tau I \epsilon \Gamma$ and ϵ_α are arbitrary, the variation of s leads to (6.3):

$$D_\tau I = 0 \tag{6.18}$$

The Euler-Lagrange equation for the variation of z^μ can be obtained from

$$\frac{d}{d\tau} \frac{\partial L}{\partial \dot{z}_a} = m \frac{d}{d\tau} \left[\frac{\dot{z}_a}{(-\dot{z}^2)^{\frac{1}{2}}} \right] - e \frac{d}{d\tau} \text{Tr}(IA^a)$$

$$= \frac{\partial L}{\partial z_a} = - e \text{Tr}(I\partial^a A^b) \dot{z}_b \tag{6.19}$$

In view of (6.18), we thus find (6.1):

$$m \frac{d}{d\tau} \left[\frac{\dot{z}^a}{(-\dot{z}^2)^{\frac{1}{2}}} \right] = -e \text{Tr}(I\mathscr{F}^{ab}) \dot{z}_b \tag{6.20}$$

The variation of A_a gives (6.2) in a standard way. Note that for this variation, the relevant term in the interaction has the conventional form

$$-e\dot{z}^a I_\alpha A^\alpha_a(z) \tag{6.21}$$

It is helpful to understand the form of L when the gauge group is U(1) [$s = e^{i\psi}$ where ψ is a real valued function]. Then we can treat K as a constant number and L differs

from the usual electromagnetic Lagrangian by a term proportional to $d\psi/d\tau$. Since the latter is the time derivative
of a function, we thus see that for the U(1) gauge group,
L is equivalent to the usual Lagrangian.

6.4 Gauge Properties of L

The Lagrangian L is invariant under the usual Yang-
Mills gauge transformation. Thus if $h(x) \in \Gamma$, it is invariant under

$$A_a(x) \rightarrow h(x)A_a(x)h(x)^{-1} + \frac{i}{e} h(x)\partial_a h(x)^{-1}$$

$$s(\tau) \rightarrow h[z(\tau)]s(\tau) \tag{6.22}$$

It is also weakly invariant under a novel gauge group.
The latter acts only on s and not on A_μ. It depends in
general on the nature of K. We thus explain it under two
headings: (A) the generic case and (B) the nongeneric case.
In the discussion which follows we assume that $K \neq 0$.

A. The Generic Case

Let $H = \{g\}$ denote all elements in Γ with the property

$$gKg^{-1} = K \tag{6.23}$$

Thus H is the stability group of K under the adjoint action.

In the generic case, the Lie algebra $\underline{H}$ corresponding
to H is just the Cartan subalgebra containing K. If $\underline{C}$ is
an a priori chosen Cartan subalgebra, then in this case, there is
a $t \in \Gamma$ such that

$$t \underline{H} t^{-1} = \underline{C} \tag{6.24}$$

For example, if G = SU(2), Γ is its two-dimensional irreducible
representation and K = σ_3, then H = U(1) = $\{e^{i(\sigma_3/2)\theta}\}$. On the other
hand, if Γ is the adjoint representation of SU(3) [so that
Γ = SU(3)/Z$_3$] and K = I$_3$, then $\underline{H}$ is spanned by I$_3$ and Y.
[The notation is standard.]

It can be shown that "most" K are of this sort. The
closure of the set of such K is all of the Lie algebra Γ
Ref. [40]. Note that for the generic case the group
H and the Lie algebra $\underline{H}$ are Abelian.

Under the gauge transformtion

$$s \to sg , \qquad g \in H \tag{6.25}$$

with s and g τ-dependent, we find

$$L \to L - i \, \mathrm{Tr} \, Kg^{-1} \dot{g} \tag{6.26}$$

The extra term is the time derivative of a function since H is Abelian. For instance, we can choose a basis K, L_α ($\alpha = 1,2,\ldots,k$) for $\underline{H}$ such that

$$\text{Tr } K L_\alpha = 0 \tag{6.27}$$

Then we can write

$$g = e^{i\theta K} e^{i\theta_\alpha L_\alpha} \tag{6.28}$$

For this form of g,

$$-i \text{ Tr } Kg^{-1}\dot{g} = \text{Tr } K^2 \dot{\theta} \tag{6.29}$$

in view of (6.27). Thus L is weakly invariant under H. The principal fiber bundle structure relevant to L is

$$H \rightarrow \Gamma \rightarrow \Gamma/H \tag{6.30}$$

where $\Gamma/H = \{sH\}$ is the space of left cosets. Thus Γ and Γ/H are the bundle and base spaces and H is the structure group.

These principal fiber bundles are never trivial. For instance, if Γ is the defining representation of $SU(2)$ and $H = U(1) = \{e^{i(\sigma_3/2)\theta}\}$ we get the Hopf fibration of the two sphere. The nontriviality of the bundle can also be seen in general. Since H, being Abelian is the product of U(1)´s (modulo perhaps a discrete group), it is infinitely connected. But Γ, being the representation of a simple compact Lie group, is finitely connected. Thus Γ ≠ Γ/HxH.

It follows that it is impossible to fix the gauge globally in this problem. However L is <u>invariant</u> under gauge transformations of the form $\{e^{i\theta_\alpha L_\alpha}\}$ [cf. equations (6.28) and (6.29)]. Thus the corresponding gauge degrees of freedom can be eliminated and L can be written in terms of a configuration space $\Gamma/\{e^{i\theta_\alpha L_\alpha}\}$. *) Since L is only weakly invariant if θ ≠ 0, the gauge degree of freedom for the U(1) gauge group $\{e^{i\theta K}\}$ cannot be so eliminated.

---

*) We note here the possibility of a topological problem which can prevent the elimination of the gauge degrees of freedom associated with $\{e^{i\theta_\alpha L_\alpha}\}$. It can occur that the ratios of the eigenvalues of K are not all rational. Then $\{e^{i\theta K}\}$ is isomorphic to the noncompact group of translations on $\mathbb{R}^1$. The topology of the latter is incompatible with the topology of the compact H. Thus, in this case, the decomposition g = $e^{i\theta K} e^{i\theta_\alpha L_\alpha}$ is incompatible with the topology of H and we cannot eliminate these gauge degrees of freedom in a smooth way.

Note that since Γ is a faithful representation of G, we can replace Γ by G in much of the preceding discussion.

The Nongeneric Case

The nonzero elements in the complement of the generic K's in $\underline{\Gamma}$ constitute the nongeneric K's [11]. The stability group

$$H = \{g\epsilon\Gamma | gKg^{-1} = K\} \tag{6.31}$$

for a nongeneric K is larger than that generated by the Cartan subalgebra containing K.

For example, if $G = SU(3)$ and $K = Y$, then $H = U(2)$. A basis for H is I_1, I_2, I_3, Y. There are no nongeneric elements for $SU(2)$.

Let K be nongeneric with stability group H. We can still choose a basis K, L_α ($\alpha = 1,2,\ldots,k$) for $\underline{H}$ with the property (6.27). Now

$$\text{Tr } K \, [L_\alpha, L_\beta] = \text{Tr } L_\alpha \, [L_\beta, K] \tag{6.32}$$
$$= 0$$

If we write

$$[L_\alpha, L_\beta] = d_{\alpha\beta\gamma}L_\gamma + \xi K \tag{6.33}$$

it follows that

$$\xi \, \text{Tr} \, K^2 = 0 \tag{6.34}$$

But $\text{Tr} \, K^2 = \text{Tr} \, K^\dagger K > 0$. Thus $\xi = 0$. The conclusion is that $\underline{H}$ is the direct sum of two Lie algebras:

$$\underline{H} = \underline{H}_0 + \underline{H}_1 \tag{6.35}$$

The algebra $\underline{H}_0$ is one dimensional and is spanned by K. The algebra $\underline{H}_1$ has a basis L_α ($\alpha = 1, 2, \ldots, k$). For the SU(3) example, $\underline{H}_0 = \underline{U}(1)$ [with basis Y] and $\underline{H}_1 = \underline{SU(2)}$ with basis I_1, I_2, I_3.

A general gauge transformation is of the form

$$g = e^{i\theta K} e^{i\theta_\alpha L_\alpha} \tag{6.36}$$

Under $s \to sg$

$$L \to L - i \, \text{Tr} \, Kg^{-1}\dot{g}$$

$$= L + \text{Tr} \, K^2 \dot{\theta} - i \, \text{Tr} \, K\{e^{-i\theta_\alpha L_\alpha} \frac{d}{d\tau} e^{i\theta_\alpha L_\alpha}\} \tag{6.37}$$

Since $\underline{H}_1$ is a Lie algebra, the term within the parentheses is in $\underline{H}_1$. Hence the last term is zero by (6.27). It

follows that L is weakly invariant under gauge transformations due to H. The gauge group in this case is in general
non-Abelian.

The principal fiber bundle structure is

$$H \to \Gamma \to \Gamma/H \qquad\qquad (6.38)$$

as in the generic case. It is nontrivial because H is infinitely connected. The latter statement is proved as
follows: Let H_0 and H_1 be the groups associated with $\underline{H}_0$ and
$\underline{H}_1$. Then the group for $\underline{H}_0 + \underline{H}_1$ is $H = H_0 \otimes H_1$, possibly
modulo some discrete finite group. Thus H is infinitely
connected. For example, if $\Gamma = SU(3)$ and $K = Y$, H is $U(2)$
which is infinitely connected. [H is not $SU(2) \times U(1)$,
but $SU(2) \times U(1)/Z_2$.]

As in the generic case, L is invariant under H_1. Thus
the H_1 gauge freedom can be eliminated and L can be written
as a function on Γ/H_1. After this partial elimination of
gauge freedom, there still remains the H_0 gauge freedom
and the principal fiber bundle structure

$$H_0 \to \Gamma/H_1 \to \Gamma/H \qquad\qquad (6.39)$$

This remaining gauge freedom cannot be eliminated.

The gauge group can thus be reduced to $U(1)$ in L in

both the generic and nongeneric case.* In fact, in almost
all our examples from particle mechanics, the gauge group
is either U(1) or can be reduced to U(1) by a process similar
to the one above. In chapter 10 we prove a general theorem
which shows that under certain assumptions nothing more involved than U(1) bundles need appear in mechanics. That
is, we show that a global Lagrangian can always be found
by enlarging the space of degrees of freedom appropriate
to the equations of motion to at most a U(1) bundle on this
space.

6.5 An Application: Scattering Off a 't Hooft-Polyakov Monopole

As an aside, we now illustrate how one can apply the
preceding formalism to probe a specific Yang-Mills field
configuration. The field configuration of interest is that
of the 't Hooft-Polyakov monopole solution [41]. When placed
at large distances from the monopole's center, the Yang-
Mills particle is known to behave similarly to that of an
electric charge in a Dirac monopole field [42]. This follows quite
simply through the use of the Lagrangian (6.9), as is shown below.

In this example $G = SU(2)$ and we may set $T(\alpha) = 1/\sqrt{2}\ \sigma_\alpha$.

---

*With the possible exception noted in the previous footnote.

At large distances from the center of the 't Hooft-Polyakov
monopole the potentials A_μ take the form

$$A_i(x) = \frac{1}{2e|x|^2} \varepsilon_{\alpha i j} \, x_j \, \sigma_\alpha \, ,$$

$$A_0(x) = 0 \, , \qquad |x|^2 = x_i x_i \, , \qquad i = 1,2,3 \qquad (6.40)$$

Here we shall restrict the discussion to nonrelativistic
particles. Upon substituting into L we obtain

$$L = \tfrac{1}{2} m \dot{z}_i^2 - i \, \text{Tr} \, Ks^{-1} \dot{s} - \tfrac{i}{4} \, \text{Tr} \, Ks^{-1} [\hat{z}, \dot{\hat{z}}] s$$

$$\hat{z} = \frac{z_i \sigma_i}{r} \qquad\qquad r = |z| \qquad\qquad (6.41)$$

In analogy with Chapter 4, let us write

$$\hat{z} = t\sigma_3 t^{-1} \qquad\qquad\qquad (6.42)$$

where $t \in \Gamma$ will be regarded as a dynamical variable defining $\hat{z}$. Notice that the dynamics of this system will not
be altered if we make the replacement*

---

*Variations of s can be implemented through variations
of u. They can also be implemented through variations of t,
which will simultaneously rotate the particle and its isospin $\vec{I}$.
Clearly, the above two variations are equivalent to varying
s and t independently.

$$s = tu , \qquad u \in \Gamma \tag{6.43}$$

in (6.41). Thus an equivalent Lagrangian for this system is

$$L' = \tfrac{1}{2}m\dot{z}_i^2 - i \, \mathrm{Tr} \, K(tu)^{-1} \frac{d}{d\tau} (tu) - \tfrac{1}{4} \, \mathrm{Tr} \, K(tu)^{-1} [\hat{z}, \dot{\hat{z}}] (tu) \tag{6.44}$$

$$= \tfrac{1}{2}m\dot{r}^2 + \tfrac{1}{2}mr^2 \, \mathrm{Tr} \, \dot{\hat{z}}^2 - \tfrac{1}{2} \, \mathrm{Tr}(I\sigma_3) \, \mathrm{Tr}(\sigma_3 t^{-1}\dot{t}) - i \, \mathrm{Tr} \, Ku^{-1}\dot{u} \tag{6.45}$$

Let us now take up the equations of motion. Variations of the coordinate z_i will be performed through variations of r and t [cf. Chapter 4]. Variations of u yield

$$\dot{I} - \tfrac{1}{2} \, \mathrm{Tr} \, (\sigma_3 t^{-1}\dot{t}) \, [I, \sigma_3] = 0 \tag{6.46}$$

The isospin vector thus precesses around the third direction in internal space. The precessional frequency depends on the position of the particle through the variable t. By taking the trace of (6.46) with σ_3, we find

$$\mathrm{Tr} \, \sigma_3 I = -2n = \text{constant} \tag{6.47}$$

The remaining equations of motion are obtained from variations in the first three terms in (6.45). Notice that the first three terms are identical to the Lagrangian (4.14) describing

a charged particle in a Dirac monopole field [with the
assignment (6.47)]. Thus the Yang-Mills particle behaves
as a charged particle in a Dirac monopole field with additional internal dynamics given by equation (6.46). Here
the correspondence is

$$\frac{eg}{2\pi} \leftrightarrow - \text{Tr}\sigma_3 I \tag{6.48}$$

Unlike the charge-monopole system of Chapter 4, n is not a
fixed number in the Lagrangian, but rather a dynamical quantity which obeys the inequality:

$$n^2 \leq \tfrac{1}{2} \text{Tr } I^2 \tag{6.49}$$

We thus expect that in the quantum mechanical system for the
particle, a spectrum in n will appear, consistent with the
inequality (6.49).

7. Kaluza-Klein Theory

The unified field theory of Kaluza and Klein [43] has been experiencing a revival of interest since the development of gauge field theories in elementary particle physics. Here the dynamical fields, denoted collectively by ψ, depend both on a space - time coordinate x and a group element s, i. e. $\psi = \psi\,(x,s)$. If $M^4 = \{x\}$ is the Minkowski space and $G = \{s\}$ denotes the internal symmetry group, the fields are thus defined on the principial fibre bundle $M^4 \times G$.[44,45].

In this section, we will discuss such theories in the context of particle mechanics. For extended objects [46], the Kaluza-Klein formalism can be generalized in a straightforward manner [47].

7.1 Kaluza-Klein Description of Point Particles

The conventional description of the Kaluza-Klein formalism is as follows. Let x^a denote the space-time coordinate of the particle. Let G={s} be a semi-simple, compact Lie group represented by unitary matrices. Here we wish to use G to describe the internal degrees of freedom of the particle. The natural metric to be used on $M^4 \times G$ is a combination of the invariant line element on M^4 and the left invariant metric on G [44].

The Lagrangian is chosen to be

$$L = -m(\ -\dot{x}^2 - \lambda Tr(s^{-1}\dot{s}s^{-1}\dot{s})\ )^{1/2} \qquad (7.1)$$

Here m and λ are constants, and $x^a = x^a(\tau)$, $s = s(\tau)$. Geo-
metrically, this Lagrangian has the following meaning. Let
us enlarge the Minkowski space M^4 to $M^4 \times G$ and regard the latter
as the configuration space. Recall that the Lagrangian for a
free particle possessing no internal symmetries is proportional
to the invariant length in M^4. Similarly, the Lagrangian (7.1)
is proportional to the invariant length on $M^4 \times G$.

The system given by (7.1) has the following properties:

i) The states in the quantum system described by L belong to a
reducible representation of G. This differs from the quantum
system for the Yang-Mills particle described in Chap.6 (Cf. Sec.8.4).

ii) The square of the momentum, p^2, depends on the quadratic
Casimir operator. This leads to a mass spectrum for the particle.

Regarding i) we note that quantum mechanical Hilbert space
carries the regular representation (see Sec.8.5). Thus the
multiplicity of an irreducible representation is equal to its
dimension by the theorem of Peter and Weyl [10].

We can show ii) by computing the operator p_a, which generates
translations, and the internal generators I_α for L, and showing
that they are algebraically related. Thus we find

$$p_a = \frac{\partial L}{\partial \dot{x}^a} = \frac{m^2 \dot{x}_a}{L} \qquad (7.2)$$

The generators I_α can be found by examing the variation

$$\delta s = i \varepsilon_\alpha T(\alpha) s \qquad (7.3)$$

where $T(\alpha)$ are the hermitian generators of G, which fulfill

$$\text{Tr}(T(\alpha)T(\beta)) = \delta_{\alpha\beta} \tag{7.4}$$

It follows that

$$\delta L = -i\dot{\epsilon}_\alpha \frac{m^2\lambda}{L} \text{Tr}(T(\alpha)\dot{s}s^{-1}) \tag{7.5}$$

Thus the quantities

$$I_\alpha = i\frac{m^2\lambda}{L} \text{Tr}(T(\alpha)\dot{s}s^{-1}) \tag{7.6}$$

are conserved. In Sec.8.5 they will be shown to generate internal symmetry transformations. Now

$$p_a p^a = \frac{m^4}{L^2}\dot{x}^2 \tag{7.7}$$

and furthermore,

$$I_\alpha I_\alpha = -\frac{m^4\lambda^2}{L^2} \text{Tr}(\dot{s}s^{-1}\dot{s}s^{-1}) = -\frac{m^4\lambda^2}{L^2} \text{Tr}(s^{-1}\dot{s}s^{-1}\dot{s}) \tag{7.8}$$

where we have used the completeness of the generators, i.e.

$$T(\alpha)\text{Tr}(T(\alpha)\dot{s}s^{-1}) = \dot{s}s^{-1} \tag{7.9}$$

Hence we obtain

$$p_a p^a - \frac{1}{\lambda} I_\alpha I_\alpha = - m^2 \qquad (7.10)$$

By defining the mass M as $p^2 = M^2$, we can rewrite (7.10) as follows

$$M^2 = m^2 - \frac{1}{\lambda} I_\alpha I_\alpha \qquad (7.11)$$

If λ is less than zero, the M^2-spectrum increases with the quadratic Casimir operator. If λ is larger than zero, the mass M becomes imaginary for some value of $I_\alpha I_\alpha$, L becomes complex and the system is inconsistent.

7.2 Reformulation of the Kaluza-Klein Theory

We shall now formulate the Kaluza-Klein Lagrangian in a different way. Although the classical equations of motion for this new system are identical to those discussed in the previous section, the corresponding quantum theories differ. Unlike in the previous section, the quantum mechanical Hilbert space derived from the following Lagrangian carries an irreducible re-presentation of the group G.

The idea here is a simple generalization of the Lagrangian formalism used to describe the relativistic point particle (as discussed in detail in chapter 5). For the latter, if the mass is m and the spin is zero, the Lagrangian has the form

$$L = p_a \dot{x}^a \qquad (7.12)$$

where $p_a p^a = - m^2$. We can generalize (7.12) to

$$L = p_a \dot{x}^a + i \, TrKs^{-1}\dot{s} \tag{7.13}$$

where p_a is now defined by

$$p_a = (m^2 + 1/\lambda \, TrK^2)^{1/2} \Lambda_{a0} \tag{7.14}$$

If $K = K_\alpha T(\alpha)$ is treated as a dynamical variable we recover precisely the system discussed in the previous section, where the quantum mechanical Hilbert space carries the left regular representation of G. Here, however, K will be treated as a constant.

The equivalence of (7.14) and (7.1) at the classical level is now shown by proving that for the Lagrangian (7.13), TrK^2 is the quadratic Casimir operator of the generators of G. Now consider the variation (7.3) of s for which

$$\delta (iTrKs^{-1}\dot{s}) = - \, Tr(sKs^{-1}T(\alpha)\dot{\varepsilon}_\alpha) \tag{7.15}$$

Consequently, the following charges J_α are conserved

$$J_\alpha = Tr(T(\alpha)sK^{-1}s) \tag{7.16}$$

J_α actually form the generators of G on the quantum mechanical Hilbert space. The desired result

$$p_a p^a = -m^2 - 1/\lambda \, J_\alpha J_\alpha \tag{7.17}$$

then follows from (7.14).

Note that the system described here is equivalent to the
the description of a free Wong particle (Cf. Chap.6), except
for the constraint (7.17), which resulted from the redefinition
of momentum. The constraint (7.17) is rather arbitrary. In fact,
we can easily arrange for any mass-internal symmetry relation by
changing the function appearing in front of Λ_{ao} in eq.(7.14). In
order to see this we notice that if $C_n(J)$, $n=1,2,\ldots$, rank (G),
denotes the Casimir invariants of G, then by (7.16), $C_n(J)=C_n(K)$.
It follows that by setting

$$P_a = f(C_1(K),C_2(K),\ldots) \Lambda_{a0} \tag{7.18}$$

for a suitable function f, we can get any mass spectrum. For a
conventional formulation of theories of this kind we refer the
reader to the work by N. Mukunda et. al. [48]. Note that the
procedure of redefining momenta (or actually the mass) of a
particle was also found to be useful in introducing an anomalous
magnetic moment for a spinning particle (Cf. Sec.5.3).

7.3 Interaction with External Yang-Mills Fields

Above we have considered a non-interacting particle with
internal degrees of freedom. The incorporation of external fields
is straightforward and as a result we can obtain the Wong
equations [38]. In order to achieve this result we replace
the time derivatives of the group element s(t) in (7.1) by the
corresponding covariant derivative (6.10), i.e. we consider the
Lagrangian [47]

$$L = - m(-\dot{x}^2 - \lambda Tr(s^{-1}D_\tau s\, s^{-1}D_\tau s))^{1/2} \tag{7.19}$$

The equation of motion for the non-Abelian charges

$$I_\alpha = i \frac{m^2 \lambda}{L} \, \text{Tr}(T(\alpha)(D_\tau s) s^{-1})$$ (7.20)

is, as before, obtained by considering the variation (7.3), i.e. $\delta s = i\varepsilon_\alpha T(\alpha)s$. The analogue of the Equation (7.5) is then

$$\delta L = -i\dot{\varepsilon}_\alpha \frac{m^2 \lambda}{L} \, \text{Tr}(T(\alpha)(D_\tau s) s^{-1})$$

$$+ \varepsilon_\alpha \frac{m^2 \lambda}{L} \, \text{Tr}\left(\left[x^a A_a, T(\alpha)(D_\tau s) s^{-1} \right] \right)$$ (7.21)

We obtain the equation of motion

$$\frac{dI(\tau)}{d\tau} = i \left[x^a(\tau) A_a(x(\tau)), \, I(\tau) \right]$$ (7.22)

i.e. the Equation (6.18).

The Euler-Lagrange equation for $x^a(\tau)$ is

$$\frac{d}{d\tau}\left(\frac{\partial L}{\partial \dot{x}_a} \right) = m^2 \frac{d}{d\tau}\left(\frac{\dot{x}^a}{L} \right) - e\frac{d}{d\tau}(IA^a)$$

$$= -e\text{Tr}(I\partial^a A^b)\dot{x}_b$$ (7.23)

as in the derivation of (6.19). By choosing the parameter τ in such a way that $L = m$, we obtain the Lorentz-Maxwell-Wong equation (6.1) in the proper time gauge.

Finally, we notice that in the presence of an external field the mass-internal symmetry relation (7.10) is changed to

$$(p_a + eI_\alpha A_a^\alpha)(p^a + eI_\alpha A^{\alpha,a}) - \frac{1}{\lambda} I_\alpha I_\alpha = -m^2$$ (7.24)

8. The Canonical Formalism and Quantization

In this section we carry out the canonical quantization for the various systems discussed in the previous chapters. Since all the Lagrangians presented here are singular, i.e. there exist constraints amongst the corresponding phase space variables, we will rely on Dirac's quantization procedure. For extensive reviews on this procedure, see Ref. 2 .

A common feature of all the systems presented here is that elements of a group G appear as dynamical variables. A method of treating group elements for setting up the canonical formalism was given in Ref. 2 and 11. We recall it below.

Let $s \epsilon$ G be parametrized by a set of variables $\xi =$ $= (\xi_1, \xi_2, \ldots, \xi_n)$ so that $s = s(\xi)$. (The functional form of s (ξ) will not be important for us). We can then regard the Lagrangian as a functional of ξ and $\dot{\xi}$ (as well as any other configuration space variables present in the system).

We first note a preliminary identity. Let us define a set of functions $f(\epsilon) = (f_1(\epsilon), f_2(\epsilon), \ldots f_n(\epsilon))$, $\epsilon = (\epsilon_1, \epsilon_2, \ldots \epsilon_n)$, by

$$e^{i T(\alpha) \epsilon_\alpha} s(\xi) = s[f(\epsilon)], \qquad f(0) = \xi \qquad (8.1)$$

where $T(\alpha)$ are the group generators with

$$[T(\alpha), T(\beta)] = i c_{\alpha\beta\gamma} T(\gamma) \qquad (8.2)$$

Differentiating (8.1) with respect to ϵ_α and setting $\epsilon = 0$, we find

$$i \, T(\alpha) \, s(\xi) = \frac{\partial s(\xi)}{\partial \xi_\beta} \, N_{\beta\alpha}(\xi) \tag{8.3}$$

$$N_{\beta\alpha}(\xi) = \frac{\partial f_\beta(\epsilon)}{\partial \epsilon_\alpha}\Bigg|_{\epsilon=0} \tag{8.4}$$

Here det $N \neq 0$, for if not, there exist x_α, not all zero, such that $N_{\rho\sigma}x_\sigma = 0$. By (8.3), $x_\sigma T(\sigma)s(\xi)$, and hence $x_\sigma T(\sigma) = 0$. But this contradicts the linear independence of the $T(\alpha)$'s.

Now the coordinates ξ_α and their conjugate momenta π_α fulfill the Poisson bracket (PB) relations

$$\{\xi_\alpha, \xi_\beta\} = \{\pi_\alpha, \pi_\beta\} = 0$$
$$\{\xi_\alpha, \pi_\beta\} = \delta_{\alpha\beta} \tag{8.5}$$

Since N is non-singular we can replace the phase space variables π_α by t_α,

$$t_\alpha = -\pi_\beta \, N_{\beta\alpha} \tag{8.6}$$

From (8.3) and (8.5) it follows that

$$\{t_\alpha, s\} = i \, T(\alpha) \, s \tag{8.7a}$$

$$\{t_\alpha, s^{-1}\} = -i \, s^{-1} \, T(\alpha) \tag{8.7b}$$

$$\{t_\alpha, t_\beta\} = c_{\alpha\beta\gamma} \, t_\gamma \tag{8.7c}$$

To prove (8.7c) note that from the Jacobi identity

$$\{\{t_\alpha, t_\beta\}, s\} = -\{\{t_\beta, s\}, t_\alpha\} - \{\{s, t_\alpha\}, t_\beta\} \tag{8.8}$$

$$= i \, c_{\alpha\beta\gamma} \, T(\gamma) \, s$$

from (8.7a). Thus

$$\{t_\alpha, t_\beta\} = c_{\alpha\beta\gamma} \, t_\gamma + F \tag{8.9}$$

where $\{F, s(\xi)\} = 0$. Consequently F is independent of the π's. Substituting $\pi_\alpha = 0$ in (8.9), we find F=0. This proves (8.7c).

The PB's (8.7) involving t_α and s are simple and do not require a particular parametrization for $s(\xi)$. We therefore find it convenient to use these variables in canonically quantizing the systems below.

8.1 Non-Relativistic Spinning Particles

Here we show how the Hamiltonian description for a spinning particle (eqs. (3.1)-(3.6)) is obtained from the Lagrangian (3.19) [(3.26)]. Now G is $SU(2) = \{s\}$ and $T(i) = \frac{1}{2} \sigma_i$. The phase space coordinates are x_i, p_i, s and t_i, where p_i is canonically conjugate to x_i. From (3.19) [(3.26)] we obtain the following primary constraint:

$$\phi_i = t_i - S_i \approx 0 \tag{8.10}$$

where S_α is defined in (3.18). From (8.7),

$$\{\phi_i, \phi_j\} = \varepsilon_{ijk} (\phi_k - S_k) \tag{8.11}$$

Applying Dirac's procedure, the following Hamiltonian is obtained from the Lagrangian (3.19):

$$H = \frac{p^2}{2m} + \phi_i \, \eta_i \tag{8.12}$$

η_i are Lagrange multiples. From the requirement that

$$\{\phi_i, H\} = 0 \qquad (8.13)$$

on the reduced phase space, we find that there exist no secondary constraints. Instead, we obtain conditions on η_α,

$$\varepsilon_{ibc}\eta_b t_c = 0 \qquad (8.14)$$

Since those $\eta_i(=\eta_i^{(t)})$ in a direction parallel to t_i are left arbitrary, only those variables which have a weakly zero PB's with $\eta_i^{(t)}\psi_i$ have a well defined time evolution. Only such variables are of physical interest. We will call them observables. Of course, x_i and p_i are observables. In addition, so are t_i and S_i. This follows from

$$\{t_i, \phi_j\} = \varepsilon_{ijk}\phi_k \qquad (8.15)$$

$$\{s, \eta_i^{(t)}\phi_i\} = -\frac{i}{2}\eta_i^{(t)}\sigma_i s \approx \frac{i}{2}s\sigma_3 \qquad (8.16)$$

Eq.(8.16) (which is weakly valid) corresponds to an infinitesimal version of the U(1) gauge transformation discussed in Chap.3. Hence only those functions of s which are invariant under gauge transformations (3.35) are also observables. But these are precisely S_i (or functions thereof). However, we can eliminate S_i by applying the constraints. Thus a complete set of observables on the reduced phase space are

$$x_i, \quad p_i \quad \text{and } t_i \qquad (8.17)$$

since S_i can be eliminated via the constraints. In so doing note that

$$t_i t_i = \lambda^2 \qquad\qquad (8.18)$$

It remains to compute the Dirac Brackets (DB's) for the variables (8.17). But these are identical to the corresponding PB's since all variables (8.17) have (weakly) zero PB with the constraints (Cf. eq. (8.15)). Consequently, we have recovered the Hamiltonian description for a non-relativistic spinning particle (Eqs. (3.1)-(3.5)). (Note that instead of eliminating S_i via the constraints, we could have eliminated t_i. In this case the DB's involving S_i do differ from the corresponding PB's. It can be shown that DB's for two S_i's are given by (3.4). Consequently, both procedures are equivalent.) It is straighforward to repeat the above analysis in the case where a spinning particle with magnetic moment μ is placed in an external magnetic field.

In passing to the quantum mechanical system, as usual, we replace the Poisson bracket by (-i) times the commutator bracket. Now the particular representation which occurs in the quantum theory is determined by λ (Cf. eq. (8.18)). This implies that (a) only one irreducible representation (IRR) appears in the theory, and (b) quantization is possible only if λ^2 is restricted to having the values

$$t^2 = 1(1+1) \ , \ 1=0, \ 1/2, \ 1, \ \ldots \qquad\qquad (8.19)$$

Note (b) is similar to the Dirac charge quantization condition which occurs in magnetic monopole theory.

8.2 Magnetic Monopoles

The canonical quantization for the magnetic monopole theory proceeds in a similar fashion to the proceeding section. The essential difference is due to equation (4.12), which constrains the configuration space variables for the monopole. Consequently, the independent phase space coordinates now consist of r, p_r, s and t_i, where p_r is canonically conjugate to r. From eq. (4.14) we find only one primary constraint,

$$\phi = \hat{x}_i t_i - n \approx 0 \tag{8.20}$$

where $\hat{x}_i$ is defined in eq. (4.12). Computing the Hamiltonian we have

$$H = \frac{p_r^2}{2m} + \frac{1}{2mr^2}(t_i t_i - n^2) + \eta\phi \tag{8.21}$$

Here η is a Lagrange multiplier. The constraint (8.20) is rotationally invariant: $\{\phi, t_i\} = 0$. Hence the requirement that $\{\phi, H\} = 0$ on the reduced phase space leads to no secondary constraints.

As before, observables are those variables which have zero PB's with ϕ. Among them are

$$r, p_r, t_i \text{ and } \hat{x}_i \tag{8.22}$$

The latter follows from

$$\{\phi, s\} = \frac{i}{2} s\sigma_3 \tag{8.23}$$

which is analogous to (8.16). As before, this corresponds to a U(1) gauge transformation, and only those functions of s which are invariant under such transformations are observables. But these are precisely $\hat{x}_i$ (or functions thereof), so (8.22) corresponds to a complete set of observables subject to the constraint (8.20).

A representation for the quantum theory can be constructed as follows. Let us regard the wave functions as functions of r and s:

$$\psi = \psi(r,s) \tag{8.24}$$

The position coordinates are diagonal in this representation in view of (4.12). The momentum p_r acts as the usual differential operator on ψ. The operators t_i are the differential operator which represent the elements $\frac{1}{2}\sigma_i$ in the left regular representation of SU(2), i.e.,

$$\left[\exp(i\theta_i t_i)\psi\right](r,s) = \psi(r,\exp(-i\theta_i\sigma_i/2)s) \tag{8.25}$$

The constraint (8.20) is taken into account by imposing the condition

$$\hat{x}_i t_i \psi = n\psi \tag{8.26}$$

on the wave functions. In view of (8.25) and (4.12), this means

$$\psi(r,s \exp(-i\theta\sigma_3/2)) = \psi(r,s)\exp(i\theta n) \tag{8.27}$$

The scalar product of wave functions is

$$(\psi,\chi) = \int_0^\infty dr r^2 \int_{SU(2)} d\mu(s)\psi^*(r,s)\chi(r,s) \tag{8.28}$$

where $d\mu$ is the invariant Haar measure on SU(2).

Let $\{D^j(s)\}$ be the representation of SU(2) with angular momentum j. Wave functions ψ with finite norm have the expansion [Ref.10]

$$\psi(r,s) = \sum_j \sum_{\rho,\sigma} \alpha^j_{\rho\sigma}(r) D^j_{\rho\sigma}(s) \qquad (8.29)$$

Here $D^j_{\rho\sigma}$ are the matrix elements of D^j in the conventional basis with the third component of angular momentum diagonal.

The constraint (8.27) means that in eq.(8.29), only those $\alpha^j_{\rho\sigma}$ with $\sigma = -n$ are non-zero. Thus

$$\{D^j_{\rho,-n}\} \quad , \text{ fixed n} \qquad (8.30)$$

is a basis for expansions of the form (8.29). Since σ is necessarily integral or half-integral, we have the Dirac quantization condition

$$2n = \text{integer} \qquad (8.31)$$

In (8.30), j and ρ are half-integral if 2n is odd and integral if 2n is even.

The quantum mechanics outlined here is essentially equivalent to conventional treatments.

8.3 Relativistic Spinning Particles

In this section we shall only be concerned with free relativistic spinning particles.

The group G is now the connected component of the Lorentz group $L^\uparrow_+ = \{|\Lambda^a{}_b|\}$ with generators $\sigma_{ab} = |(\sigma_{ab})^c{}_d|$ defined in eq.(5.5). Here (8.2) reads

$$[\sigma_{ab}, \sigma_{cd}] = i(-\eta_{bc}\sigma_{ad} + \eta_{bd}\sigma_{ac} + \eta_{ac}\sigma_{bd} - \eta_{da}\sigma_{bc}) \quad (8.32)$$

In addition (8.7c) and (8.7a) are replaced by

$$\{t_{ab}, t_{cd}\} = -\eta_{bc}t_{ad} + \eta_{bd}t_{ac} + \eta_{ac}t_{bd} - \eta_{da}t_{bc} \quad (8.33)$$

$$\{t_{ab}, \Lambda\} = i\sigma_{ab}\Lambda \quad (8.34)$$

A. Spinless particles

For simplicity, we begin with the case where the spin is absent, i.e. $\lambda=0$ in eq. (5.11). The phase space coordinates are given by z_a, π_a, Λ and t_{ab}, where π_a is canonically conjugate to z_a. The primary constraints are

$$\phi_{ab} = t_{ab} \approx 0 \quad (8.35)$$

$$\theta_a = p_a - \pi_a \approx 0 \quad (8.36)$$

where p_a is defined in (5.3). Equation (8.35) follows because there are no time derivatives of Λ appearing in the Lagrangian. The constraints obey the PB algebra:

$$\{\theta_a, \theta_b\} = 0 \quad (8.37)$$

$$\{\phi_{ab}, \theta_c\} = i(\sigma_{ab})_{cd}p^d \quad (8.38)$$

along with (8.33).

Because of the reparametrization symmetry of the Lagrangian, the Hamiltonian consists soley of the constraints (for a discussion of this issue, see, for example, Ref.3), i.e.

$$H = \rho^{ab}\phi_{ab} + \kappa^a\theta_a \quad ,$$
(8.39)

where ρ^{ab} and κ^a are Lagrange multipliers. Once again, there are no secondary constraints. Instead ρ_{ab} and κ_a are restricted by

$$\rho^{ab}P_b = 0$$
(8.40)

$$\kappa_a P_b - \kappa_b P_a = 0$$
(8.41)

on the reduced phase space. In deriving (8.40) and (8.41) we have used the representation for σ_{ab} given by eq.(5.5). Equations (8.40) and (8.41) imply that

$$\rho_{ia} = \epsilon_{ijk}r_j\Lambda^{-1}_{ka} \quad , \quad i,j,k = 1,2,3$$
(8.42)

$$\kappa_a = kp_a$$
(8.43)

where r_i, i=1,2,3 and k are undetermined constants. This, in turn, implies that four linearly independent combinations of (8.35) and (8.36) form first class constraints, namely

$$\Phi_i \equiv \epsilon_{ijk}\phi_{ja}\Lambda^{-1}_{ka}$$
(8.44)

$$\Phi_0 \equiv \theta_a\Lambda^a_0$$
(8.45)

Observables, by definition, have zero PB's with ϕ_a. Among them are

$$\pi_a \quad \text{and} \quad J_{ab} = z_a\pi_b - z_b\pi_a$$
(8.46)

where we have applied the constraints. Additional observables can be formed from P_a and t_{ab}, however, these degrees of freedom can be eliminated via the constraints (8.35) and (8.36). There exist six independent observables amongst the remaining ten degrees of freedom for the system. The former are exactly given by (8.46), since four constraints now exist on the variables π_a and J_{ab},

$$\pi^a = -m^2 \tag{8.47}$$

$$W^a = 0 \ , \quad W^a = \frac{1}{2} \epsilon^{abcd} \pi_b J_{cd} \tag{8.48}$$

(Note that eq.(8.48) yields 3 relations since $\pi_a W^a$ is identically zero). Equations (8.47) and (8.48) indicate that the above system describes a particle of mass m and spin zero.

It remains to compute the DB's for the variables (8.46). We first define J^*_{ab}

$$J^*_{ab} = J_{ab} + \phi_{ab} \tag{8.49}$$

which, along with π_a, form a complete set of first class variables. Consequently, all DB's involving J^*_{ab} and π_a are identical to the corresponding PB's. Equivalently, we can define a DB with J_{ab}, according to

$$\{J_{ab}, \ . \ \} \equiv \{J^*_{ab}, \ . \ \} \tag{8.50}$$

Using eq.(8.50), we obtain the usual Poincaré algebra for π_a and J_{ab},

$$\{\pi_a, \pi_b\}^* = 0$$

$$\{J_{ab}, \pi_c\}^* = \eta_{ac}\pi_b - \eta_{bc}\pi_a \qquad (8.51)$$

$$\{J_{ab}, J_{cd}\}^* = \eta_{ac}J_{bd} + \eta_{bd}J_{ac} + \eta_{ad}J_{cb} + \eta_{bc}J_{da}$$

Note that equations (8.47) and (8.48) lie in the center of the algebra generated by J_{ab} and π_a. So if desired, one can eliminate redundant variables from J_{ab} and π_a, by hand, without conflict with their DB's (8.51).

Since the Hamiltonian is simply a linear combination of the constraints (8.44) and (8.45), it generates no time evolution for J_{ab} and π_a. If desired, we can declare that π_0 generates time translations. Also we can identify

$$x_i = \frac{1}{\pi_0} J_{i0} \quad , \quad i=1,2,3 \qquad (8.52)$$

as the space coordinate of the particle. It fulfills

$$\{x_i, \pi_j\}^* = \delta_{ij} \qquad (8.53)$$

$$\{x_i, x_j\}^* = 0 \qquad (8.54)$$

In proving eq.(8.54), a direct computation yields

$$\{x_i, x_j\}^* = \frac{1}{\pi_0^2} \left(- J_{ij} + \frac{1}{\pi_0}(J_{i0}\pi_j - J_{j0}\pi_i) \right) \qquad (8.55)$$

The result is then obtained after applying the definition for J_{ab} (Cf. eq.(8.46)).

B. Spinning particles

In this case, we consider non-zero values for λ in eq. (5.11).

Here, eq. (8.35) is replaced by

$$\phi_{ab} = t_{ab} - S_{ab} \approx 0 \tag{8.56}$$

where S_{ab} is given by eq. (5.4). Equation (8.56), along with (8.36), form the primary constraints for this system. Their PB's are given by (8.37), (8.38) and

$$
\begin{aligned}
\{\phi_{ab}, \phi_{cd}\} &= n_{bc}(\phi_{da} - S_{da}) - n_{ad}(\phi_{bc} - S_{bc}) \\
&+ n_{ac}(\phi_{bd} - S_{bd}) - n_{bd}(\phi_{ca} - S_{ca})
\end{aligned}
\tag{8.57}
$$

The Hamiltonian, once again, consists soley of the constraints, i.e. eq. (8.39). Again there are no secondary constraints, and, instead, the Lagrange multipliers are restricted by (8.40) and

$$\kappa_a P_b - \kappa_b P_a = 2(S_{ac}\rho^c{}_b - S_{bc}\rho^c{}_a) \tag{8.58}$$

on the reduced phase space. After applying the definitions for S_{ab} and P_a (Cf. eqs. (5.3) and (5.4)), we find

$$
\begin{aligned}
0 &= \frac{m}{2\lambda}(\tilde{\kappa}_a n_{b0} + n_{a1}\tilde{\rho}_{b2} - n_{a2}\tilde{\rho}_{b1}) - (a \rightleftharpoons b) \quad , \\
\tilde{\kappa} &= \Lambda^{-1}\rho \ , \quad \tilde{\rho} = \Lambda^{-1}\rho\Lambda
\end{aligned}
\tag{8.59}
$$

Equations (8.40) and (8.59), along with

$$\tilde{\rho}_{ab} = -\tilde{\rho}_{ab} \tag{8.60}$$

imply that all components of $\tilde{\kappa}$ and $\tilde{\rho}$ vanish except for $\tilde{\kappa}_o$ and $\tilde{\rho}_{12}$. Consequently, there are two first class constraints: Eq. (8.45) and

$$\phi_{ab} \Lambda^{a1} \Lambda^{b2} \approx 0 \tag{8.61}$$

Once again π_a and J_{ab} (Cf. eq. (8.46)) are observables for the system. However, they no longer form a complete set of observables. Since now only two first class constraints can be found, there exist a total of eight observables for the system; but there are only six independent degrees of freedom in π_a and J_{ab}. Additional observables for this system are S_{ab}. Note that there are four constraining equations on S_{ab}

$$S_{ab} \pi^b = 0 \tag{8.62}$$

$$\frac{1}{2} S_{ab} S^{ab} = \lambda^2 \tag{8.63}$$

(Eq. (8.62), which holds on the reduced phase space, contains a total of three constraints since $S_{ab} \pi^a \pi^b$ vanishes identically). Thus two independent degrees of freedom remain in S_{ab}; so π_a, J_{ab} and S_{ab} form a complete set of observables.

Alternatively, the five independent degrees of freedom in J_{ab} and S_{ab} can be expressed more compactly by

$$M_{ab} = J_{ab} + S_{ab} \tag{8.64}$$

M_{ab} contains all five degrees of freedom, since

$$W_a W^a = m^2 \lambda^2 \quad \left[W^a \equiv \frac{1}{2} \epsilon^{abcd} \pi_b M_{cd} \right] \tag{8.65}$$

is the only constraining equation on M_{ab}.

Equation (8.65) indicates that particle has a fixed spin λ. It, along with (8.47), can be used to eliminate, by hand, the redundant degrees of freedom from π_a and M_{ab}. This follows because (8.47) and (8.65) lie in the center of the algebra generated by π_a and M_{ab}. It remains to be shown that this algebra is, once again, the Poincaré algebra.

With this in mind, we define

$$M^*_{ab} = M_{ab} + \phi_{ab} = J_{ab} + t_{ab} \tag{8.66}$$

which, along with π_a, form a complete set of first class variables. DB's involving M_{ab} can then be defined by

$$\{M_{ab}, \cdot \}^* = \{M^*_{ab}, \cdot \} \tag{8.67}$$

while DB's involving π_a are equivalent to the corresponding PB's. Using (8.67) we then verify that π_a and M_{ab} generate the Poincaré algebra.

Again, if desired, we can declare that time translations are generated by π_0. The standard canonical DB's (8.53) and (8.54) are obtained after defining the space coordinate x_i according to (see Sudarshan and Mukunda, Ref.2, p.439-454)

$$x_i = \frac{1}{\pi_0}(M_{i0} - \frac{\epsilon_{ijk}\pi_j W_k}{m(m- \pi_0)}) \tag{8.68}$$

Note that eq.(8.68) does reduce to (8.52) in the limit of zero spin. In addition, spin 3-vectors $\tilde{S}_i$, with the usual brackets

$$\{\tilde{S}_i,\tilde{S}_j\}^* = \epsilon_{ijk}\tilde{S}_k \tag{8.69}$$

can be defined in terms of the Poincare generators [2]:

$$\widetilde{S}_i = -\frac{1}{m} \left(W_i + \frac{W_j \pi_j \pi_i}{\pi_0 (m - \pi_0)} \right) \tag{8.70}$$

The variables $\widetilde{S}_i$ differs from $S_i = \frac{1}{2} \epsilon_{ijk} S_{jk}$, which can be reconstructed in terms of the Poincare generators. The latter variables do not satisfy eq. (8.69).

In conclusion, the eight degrees of freedom in π_a and M_{ab} can be expressed in terms of x_i, π_i and $\widetilde{S}_i$, which have standard bracket relations. The Hamiltonian for π_0 is

$$H = \left(\pi_i^2 + m^2 \right)^{1/2} , \tag{8.71}$$

where we have chosen the positive root in elminating the constraint (8.47). In terms of the variables x_i, π_i and S_i the constraint (8.65) translates to

$$\widetilde{S}_i \widetilde{S}_i = \lambda^2 \tag{8.72}$$

This system represents the obvious generalizations of the non relativistic spinning particle system described in Secs. 3.1 and 8.1, As before, we find that only one IRR appears in the quantum theory, and quantization is possible only if λ^2 is restricted to having the values given in eq. (8.19).

8.4 Yang-Mills Particle

For simplicity we shall specialize to the case of non - relativistic particles. Consequently, we replace the first term in (6.9) by $\frac{1}{2} m \dot{x}_i^2$. Now the phase space coordinates are x_i, p_i, s, t_α. Here $s \epsilon \Gamma$, Γ being a faithful unitary representation of an arbitrary compact connected Lie group G.

The Hamiltonian for this system is

$$H = \frac{1}{2m} (p_i - eA_i^\alpha t_\alpha)^2 + eA_0^\alpha t_\alpha + \eta_\alpha \phi_\alpha \qquad (8.73)$$

where the primary constraints are given by

$$\phi_\alpha = I_\alpha - t_\alpha \approx 0 \qquad (8.74)$$

In deriving (8.73) we have used the constraints to rearrange

terms. As usual, there are no secondary constraints and the

η's are restricted by

$$c_{\sigma\rho\lambda} \eta_\rho t_\lambda = 0 \qquad (8.75)$$

on the constrained surface. Let there be k independent

vectors $\{\eta_\rho = \eta_\rho^{(A)}, \quad A=1,2,\ldots k\}$ satisfying (8.75).

The first class constraints of the theory are

$$\phi^A = \eta_\rho^{(A)} \phi_\rho \qquad (8.76)$$

Observables have zero PB's with ϕ^A. They consist of

$$x_i, \ p_i, \ t_\alpha \qquad (8.77)$$

Note that the I_α's are also observables. This follows from

$$\{s, \phi^A\} = -\eta(A)s \qquad , \qquad (8.78)$$

where $\eta(A) = \eta_\alpha^{(A)} T(\alpha)$ generate the stability group of $t = t_\alpha T(\alpha)$

under the adjoint action. This group is isomorphic to the group H

(Cf. Sec.6.4). From (8.78), only those functions of s which

are invariant under the action the little group of t are of

interest. These must be functions of I . However, the I 's

can be eliminated via the constraint. Thus we are left with

variables (8.77). Since they all have weakly zero PB with ϕ_α, all DB's involving these variables are identical to the corresponding PB's.

As in Sec.1, not all the t_α's are independent. From (6.13) and (8.74), t is constrained to lie on a certain orbit in $\underline{r}$. These orbits are labeled by the constants K_α. Using (6.13) any function of the t_α's which is a constant on the orbits can be written as a function of the K_α's. In particular, the Casimir invariants can be expressed in terms of K. For the case of G=SU(2), we are left with one constraint, which is analogous to (8.18).

The particular representation which occurs in the quantum theory is determined by the Casimir invariants, which in turn are determined by the K_α's. Once again, only one IRR appears in the quantum theory and quantization is possible only if the Casimir invariants formed out of the K_α's are restricted to a certain discrete set.

8.5 Kaluza-Klein Formulation

As was true in several other cases the Lagrangian here (Cf. eq.(7.1)) contains a reparametrization symmetry. Here we shall remove it by fixing $x^0=\tau$. The Hamiltonian for this system is

$$H = (m^2 + p_i^2 - \frac{1}{\lambda}t_\alpha^2)^{1/2} \qquad , \qquad (8.79)$$

where x_i, p_i, and t_α are the usual phase space variables. Unlike the previously discussed systems, there are no constraints on the phase space variables. This is due to the fact that

t_α can be expressed in terms of $\dot{s}s^{-1}$ (Cf.(7.6) (Here $I_\alpha = t_\alpha$)).

In the previous section $t_\alpha T(\alpha)$ was constrained to lie

on certain orbits in the Lie algebra. These orbits determined

which IRR was to appear in the quantum theory. Now there are

no constraints on the variables t_α and, consequently, all

IRR's appear in the quantum theory.

In setting up the quantum theory, we can write down wave-

funtions which are functions of s as well as x_i.

$$\psi = \psi(s,x) \tag{8.80}$$

This follows since all components $s_{\alpha\beta}$ can be simulta -

neously diagonalized. Then the t_α's are differential operators

which represent the generators $T(\alpha)$ in the left regular re-

presentation of the group. In particular,

$$\{exp(i\theta_\alpha L_\alpha)\psi\}(s,x_i) = \psi(exp(-i\theta_\alpha T(\alpha))s,x_i) \tag{8.81}$$

The scalar product with respect to which the t_α's are Hermitian

is given by

$$(\psi',\psi) = \int d\mu(s)d^3x \ \psi'^*(s,x)\psi(s,x) \tag{8.82}$$

where $d\mu(s)$ is the invariant Haar measure of the group. The

left regular representation is highly reducible. Every irre-

ducible representation occurs with a multiplicity equal to

its own dimension.

If an irreducible representation of the Kaluza-Klein

system is desired, we must deal with the formulation given

in section 7.2. As was noted earlier, the system there is
identical to that of the Yang-Mill particle with the mass
given by

$$(m^2 + \frac{1}{\lambda} \text{TrK}^2)^{1/2}$$

(8.83)

Excluding this additional requirement, the quantization of
such a system is identical to that discussed in section 8.4.

9. Pseudo-Classical Description

In the previous sections, we have seen how to describe the spin and isospin degrees of freedom of a particle in term of dynamical group elements $g(\xi) \, \epsilon \, G$. It has been pointed out however, that Grassmann variables, i.e. anticommuting c-numbers, can be utilized for the same purpose. Such a formulation is usually, referred to as pseudoclassical mechanics. Upon quantization the anticommuting c-numbers leads to certain irreducible representations of some symmetry group G. In the case of spin degrees of freedom it was shown by Volkov, Peletminskii [49] and Martin [50] that the classical Grassmann variables, are replaced by Pauli matrices after quantization. These considerations for point particles (and extended objects) have recently been discussed in much detail in the literature [51-52]. They have also been applied to internal degrees of freedom [39, 53-54].

The algebra of the anticommuting Grassmann variables can be used to extend the notion of Lie-algebras to graded Lie-algebras [5]. The notation of graded lie algebras, usually referred to as supersymmetry, has been extended to field theory leading to global and local (i.e. supergravity) supersymmetric field theories. For a review of this very dynamic field of research and for futher references see e.g. Ref. 56.

In this chapter we apply some of these concepts to the description of the systems discussed in the previous chapters.

9.1. Nonrelativistic Spinning Particles

The Lagrangian for a free, nonrelativistic spinning particle involving dynamical anticommuting Grassmannian variables, $f_a(\tau)$, is [52-53]

$$L_0 = \tfrac{1}{2}m\dot{x}_a^2 + \tfrac{i}{2}f_a\dot{f}_a \qquad (9.1)$$

The equations of motion derived form (9.1) are

$$m\ddot{x}_a = 0 \ , \ \dot{f}_a = 0 \qquad (9.2)$$

The orbital angular momentum L_a and spin S_a, as defined by

$$S_a = -\tfrac{i}{2}\varepsilon_{abc}f_bf_c \ , \qquad (9.3)$$

are both separately conserved due to (9.2).

Note that the Lagrangian (9.1) is weakly invariant under the transformation

$$x_a \to x_a - i\varepsilon f_a/\sqrt{m}$$
$$f_a \to f_a + \varepsilon\sqrt{m}\dot{x}_a \ , \qquad (9.4)$$

where ε is a c-number Grassmann parameter. We define (9.4) to be a "supersymmetry" transformation. Under (9.4),

$$L_0 \to L_0 + \tfrac{i}{2}\frac{d}{dt}(f_a\dot{x}_a\varepsilon)\sqrt{m} \qquad (9.5)$$

Thus the action $\int dtL_0$ is invariant.

Interactions can be added to the Lagrangian (9.1) in a straightforward manner, although in general, they will not be invariant under (9.4). For example, consider the interaction of a particle having magnetic moment μ (and no charge) with an external electromagnetic field $\vec{B}$. The form of the Lagrangian is then the same as in Equation (3.26), i.e.

$$L = L_0 - \mu S_a B_a \qquad (9.6)$$

A variation of the coordinate x_a leads to the equation of motion (3.31). A variation of f_a leads to

$$\dot{f}_a + \mu \epsilon_{abc} f_b B_c = 0$$

$$(9.7)$$

Using the definition (9.3) of the spin angular momentum, we obtain the spin precession equation (3.34),

$$\dot{S}_a = \mu \epsilon_{abc} B_b S_c \qquad (9.8)$$

The interaction given in (9.6) is not invariant under the super-symmetry transformation (9.4), since it transforms according to

$$-\mu S_a B_a \rightarrow -\mu S_a B_a + i\mu \sqrt{m} \epsilon_{abc} B_a \dot{x}_b f_c \qquad (9.9)$$

On the other hand, if we add the term

$$-eA_a \dot{x}_a \qquad (9.10)$$

to the interaction Lagrangian (9.6), and set

$$\mu = e/m \qquad (9.11)$$

the invariance under (9.4) is restored. Here

$$F_{ab} = \epsilon_{abc} B_c = \partial_a A_b - \partial_b A_a \qquad (9.12)$$

Next we give a superfield formulation [57-58] of the above systems. It provides us with a systematic method for constructing lagrangians which are invariant under the supersymmetry trans-formations (9.4). We define $X_a(t,\theta)$ to be a "supercoordinate", i.e. it depends an superspace parameters t and θ, the latter being a Grassmann parameter. We then identify the coefficients of the Taylor expansion of $X_a(t,\theta)$ in θ with x_a and $f_a/\sqrt{m}$, i.e.

$$X_a(t,\theta) = x_a(t) + i\theta f_a(t)/\sqrt{m} \qquad (9.13)$$

Now consider the following "supercharge" operator

$$Q = \left(i\frac{\partial}{\partial \theta} - \theta \frac{\partial}{\partial t} \right) \qquad (9.14)$$

It follows that

$$[Q,Q]_{+} = -2i\frac{\partial}{\partial t} \, , \qquad (9.15)$$

i.e. the anticommutator of two supercharges yields the energy operator. Equation (9.15) thereby expresses a general property of supersymmetry algebras [56]. Furthermore, it can be easily be verified that the supercharge Q induce translations in superspace (t,θ) according to

$$\delta X_a(t,\theta) \equiv i\varepsilon Q X_a(t,\theta) = X_a(t+i\varepsilon\theta, \theta-\varepsilon) - X_a(t,\theta) \qquad (9.16)$$

The transformation defined by the Equation (9.16) applied to the supercoordinate X_a is identical to the supersymmetry transformations (9.4) applied to x_a and f_a.

For the purpose of constructing weakly invariant Lagrangians, we now note the following:

i) Let $Y = Y(t,\theta) = y(t) + \theta\eta(t)$ be a supercoordinate, which undergoes the transformation

$$\delta Y = i\varepsilon Q Y \qquad (9.17)$$

Then $\eta(t)$ is invariant under this transformation up to a total time derivative. (This is analogous to the transformation properties of the D-term in 3+1 dimensional supersymmetry [56].)

ii) Let $P(t,\theta)$ be a (fermionic) bosonic operator which (anti-) commutes with Q. Then if the superfield Y transforms according to (9.17), so does $P(t,\theta)Y$. Examples of first order differential operators P fulfilling this property are

$$\frac{\partial}{\partial t} \, , \quad d_\theta \equiv \frac{\partial}{\partial\theta} - i\theta\frac{\partial}{\partial t} \, , \qquad (9.18)$$

the former being bosonic while the latter is fermionic

iii) If Y and Z transform according to (9.17), then so does the product YZ.

Let $L_{ox} = L_{ox}(t,\theta)$ transform under supersymmetry according to (9.17). Then the θ coefficient of L_{ox} is invariant up to a time derivative. Since we desire an invariant quantity which is bosonic, L_{ox} should be fermionic. A choice for L_{ox} which is quadratic in first order derivations of X_a is

$$L_{ox} = \frac{mi}{2} \frac{\partial}{\partial t} X_a d_\theta X_a \qquad (9.19)$$

The θ coefficient of L_{ox} can be extracted by integrating over θ and utilizing the usual rule [59]

$$\int d\theta\theta = 1 , \quad \int d\theta = 0 \qquad (9.20)$$

Applying this to (9.19), we then find

$$\int d\theta L_{ox}(t,\theta) = L_0(t) , \qquad (9.21)$$

where L_0 is the free particle Lagrangian (9.1).

Next we consider adding an interaction term to (9.19). We first take up the case of a particle interacting with a scalar (bosonic) potential $V = V(X)$. The latter transforms under supersymmetry according to (9.17). Since $V(X)$ is bosonic it must appear in $L_x = L_{ox} + L_{Ix}$ times a fermionic operator. The latter must anticommute with Q. Thus interactions like $L_{Ix} = \theta V(X)$ are excluded since they explicitly break the supersymmetry invariance. On the other hand, interactions like $L_{Ix} = d_\theta V(X)$ preserve the supersymmetry invariance. However, integrating with respect to θ leaves only a total time derivative; so no interaction results. Consequently, it appears difficult to construct a supersymmetric invariant version of a particle interacting with a scalar potential. (The latter is possible, however, for other treatments of the supersymmetric point particle, cf.Ref. [60]).

In our formalism we can quite easily write down the interaction of the particle with a vector potential $A_i = A_i(X)$. Here we simply make the replacement

$$\frac{\partial}{\partial t}X_a \rightarrow \frac{\partial}{\partial t}X_a - \frac{2e}{m}A_a(X)$$

$$d_\theta X_a \rightarrow d_\theta X_a \tag{9.22}$$

in (9.19). Expanding the total Lagrangian in θ and performing the θ-integral according to the rule (9.20),one obtains

$$L(t) = L_0(t) - ex_a A_a(x) - e/m\, S_a B_a \tag{9.23}$$

This is identical to the Lagrangian (9.6) plus (9.10),with the restriction that charge and magnetic moment are related according to (9.11).

Here we notice that the Lagrangian (9.23), or Equation (9.11), leads to the conclusion that the gyromagnetic ratio of the particle is 2. This is actually a general feature of supersymmetric point particles [see e.g. 61,24,39] . In supersymmetric field theories, where the supersymmetry is unbroken, this situation corresponds to the anomomalous magnetic moment being zero [see e.g. Ref. 61] .

In the above we have used a hermitian Grassmannian variable θ to describe the spin degrees of freedom. One can also develop a nonrelativistic supersymmetry by making use of a complex Grassmann variable. As was shown by Witten [60] and discussed by other authors [63-64] supersymmetric quantum theories can be useful for studying the non-perturbative breaking of supersymmetry.

9.2. Supersymmetric Point Particles in the Field of a Magnetic Monopole

In this section we will supersymmetrize [60] the global Lagrangian of Chapter 4. This can be achieved by applying the rules given in Section 9.1 for constructing, weakly invariant supersymmetric Lagrangians. In Chapter 4 a dynamical group element $s(t)$ entered in the construction of the global Lagrangian (4.13). We can write $s(t)$ in the form

$$s(t) = \exp(iT(a)\varepsilon_a(t)) \quad , \qquad (9.24)$$

$T(a) = \sigma_a/2$. Similarly, we can define a group element s_x on the superspace (t,θ), according to

$$s_x(t,\theta) = \exp(iT(a)\eta_a(t,\theta)) \quad , \qquad (9.25)$$

where $\eta_a(t,\theta)$ is now a "superfield". If we write $\eta_a(t,\theta) = \varepsilon_a(t) - 2\theta i \xi_a(t)$, then (9.25) can be expressed by

$$s_x(t,\theta) = (1 + \theta\xi)s(t) \quad , \qquad (9.26)$$

where $\xi = \xi_a \sigma_a$. Note that no conditions have to placed on ξ_a (other then it being an add Grassmann variable) in order that (9.26) be consistent with $s_x^\dagger s_x = 1$ and $\det s_x = 1$.

A natural extension of the Lagrangian (4.13) is

$$L_x(t,\theta) = L_{0x} - n\mathrm{Tr}(\sigma_3 s_x^\dagger d_\theta s_x) \quad , \qquad (9.27)$$

where L_{0x} is given by the Equation (9.19). We must, furthermore, generalize (4.12) (the relation between the relative coordinate x_a and the dynamical group element s) to the (t,θ) - space. This extension can now be easily achieved after constructing the following polar decomposition of the supercoordinate

$$\vec{X}(t,\theta) = \vec{x}(t) + i\theta\vec{z}/\sqrt{m} = R_x(t,\theta)\hat{X}_x(t,\theta) \quad , \qquad (9.28)$$

$$R_x(t,\theta) = r(t) + i\frac{\theta}{\sqrt{m}} \hat{x}(t) \cdot \vec{f}(t)$$
(9.29)

$$\hat{x}_x(t,\theta) = \hat{x}(t) + i\frac{\theta}{r\sqrt{m}}(\vec{f}(t) - \hat{x}(t)(\hat{x}(t) \cdot \vec{f}(t)))$$
(9.30)

In (9.29) and (9.30) $\vec{x} = r\hat{x}$. The supersymmetric generalization of the Equation (4.12) then is

$$\hat{x}_x = \hat{x}_{xa}\sigma_a = s_x\sigma_3 s_x^\dagger$$
(9.31)

The Equations (9.19), (9.26), (9.27) and (9.31) lead to the following Lagrangian

$$L = \int d\theta\, L_x(t,\theta) = \frac{1}{2}m\dot{x}_a^2 + \frac{i}{2}f_a\dot{f}_a + i n \mathrm{Tr}(\sigma_3 s^\dagger \dot{s}) +$$
$$+ 2ni\epsilon_{abc}\hat{x}_a \xi_b^\xi \xi_c^\xi$$
(9.32)

By making use of the constraint (9.31) and the explicit form of $s_x(t,\theta)$, as given by (9.26), we obtain

$$\hat{x}_x(t,\theta) = \hat{x}(t) + \theta[\xi(t),\hat{x}(t)]$$
(9.33)

where $\hat{x} = s\sigma_3 s^\dagger$. Equations (9.29) and (9.33) then leads to the following relationship

$$\epsilon_{abc}\hat{x}_a f_b f_c = 4mr^2\epsilon_{abc}\hat{x}_a \xi_b^\xi \xi_c^\xi \quad ,$$
(9.34)

i.e. the Lagrangian (9.32) can now be written in the following form

$$L = \frac{1}{2}m(\dot{r}^2 + r^2\dot{\hat{x}}^2) + \frac{i}{2}f_a\dot{f}_a + i n \mathrm{Tr}(\sigma_3 s^\dagger \dot{s}) -$$
$$- e/m\; S_a B_a$$
(9.35)

Here $\vec{B}$ is the magnetic field of the monopole, i.e.

$$B_a = \frac{g}{4\pi} x_a/r^2$$
(9.36)

and $4\pi n = eg$ (compare with section 4.1). In the expression (9.35), $\hat{x}$ is to be regarded an a function of s (cf. eg. (4.12)). For $f_a = 0$, (9.35) becomes the Lagrangian (4.13).

In order to obtain the equations of motion we consider variations of the dynamical variables r, f_a and s. The variation of r in (9.35) gives

$$m\ddot{r} = mr\dot{\hat{x}}^2 + \frac{2n}{mr^3} \vec{s} \cdot \hat{x} \tag{9.37}$$

A variation of the Grassmann variables f_a leads to a spin precession equation (cf. (9.8))

$$\dot{s}_a = \frac{e}{m} \varepsilon_{abc} B_b s_c \tag{9.38}$$

Again for variations in s, we take (cf. (3.21) and (3.22))

$$\delta s = i\varepsilon_k \sigma_k s \tag{9.39}$$

By the Equation (4.12), (9.39) will induce an infinitesimal rotation f the unit vector $\hat{x}$ as given by (4.16), i.e.

$$\delta\hat{x}_a = -2\varepsilon_{abc}\varepsilon_b \hat{x}_c \tag{9.40}$$

We therefore obtain the following result due to the variation (9.39)

$$\delta L = 2\varepsilon_a \left[\frac{d}{dt}(L_a + n\hat{x}_a) + \frac{e}{m}\varepsilon_{bac} s_b B_c\right] \tag{9.41}$$

Equations (9.38) and (9.41) can now be combined to give

$$\frac{d}{dt}(L_a + n\hat{x}_a + s_a) = 0 \quad , \tag{9.42}$$

i.e. angular momentum conservation. After a long but straightforward calculation, the Equations (9.37), (9.38) and (9.42) can be combined to yield the following equation

$$m\ddot{\vec{x}} = -e\dot{\vec{x}} \wedge \vec{B} - e/m \nabla (\vec{s} \cdot \vec{B}) \tag{9.43}$$

(cf.(3.11)), which is the equation of motion for a spinning
particle in a non-homogenous magnetic field (cf. section 3.2
and Ref. 65).[The Equation (9.43) can also be obtained directly from
the Lagrangian (9.35) by considering simultaneous variations
of s and r. Here one makes use of the relation

$$2n\dot{\hat{x}}_a \varepsilon_a = - \bar{e}(\dot{\vec{x}} \wedge \vec{B}) \cdot \delta\vec{x} \tag{9.44}$$

where $\delta\vec{x} = -2\bar{\varepsilon}\wedge\vec{x} + \delta r\hat{x}$. Equation (9.44) follows from (9.40)
and the explicit form of the magnetic monopole field (9.36).] As
expected (cf. section 9.1) the gyromagnetic ratio of the particle is 2 according to the Equation (9.38). Here we also notice
that although $\vec{L} + n\hat{x}$ is not conserved, its projection along the
$\hat{x}$-direction is. (In fact, the latter is just n.). This fact
will turn out to be important when we quantize the system
(cf. section 9.5).

9.3 The Supersymmetric Hopf Fibration

In Chapter 4 we have seen that the non-trivial $U(1)$ bundles on the two-sphere S^2 could be used to find a global
Lagrangian description of magnetic monopoles. Let us recall how
these bundles, denoted by L_M, were constructed [12, 60, 66. For
some related work see also Refs. 67-69]. In section 4.3 we
regarded $SU(2)$ as a $U(1)$ bundle over S^2, where the action of
the $U(1)$ group corresponded to the gauge transformation (4.22),
i.e.

$$s(t) \rightarrow s(t)\exp(i\sigma_3/2\alpha(t)) \tag{9.45}$$

The projection map from the $SU(2)$ bundle to the two-sphere
S^2 is given by (4.12), i.e.

$$s(t) \rightarrow s(t)\sigma_3 s^\dagger(t) = \hat{X}(t) \tag{9.46}$$

Now consider the following cyclic subgroup of SU(2)

$$Z_M = \{z_k = \exp(i\sigma_3 \frac{2\pi k}{M}) \mid k = 0,1,\ldots, M-1\} , \qquad (9.47)$$

where M is a positive integer. Z_M has an action on $s \in SU(2)$ which commutes with the projection (9.46), i.e. if

$$s \rightarrow sz_k \qquad (9.48)$$

then

$$s \rightarrow s\sigma_3 s^\dagger \rightarrow sz_k \sigma_3 (sz_k)^\dagger = \hat{X} \qquad (9.49)$$

The U(1) bundles over the two-sphere S^2 are the generated by the quotient of SU(2) with the group-action (9.48) [70]. A function f on L_M can now be regarded as a function on SU(2) which is Z_M invariant, i.e.

$$f(sz_k) = f(s) \qquad (9.50)$$

for all $z_k \in Z_M$. In view of the fact that the wave functions the charge-monopole system have the property of being $z_{|2n|}$ invariant (cf. the equation (8.27)), they can be regarded as functions on $R^1 x Z_{|2n|}$. In this sense, there is a topolozical interpretation of the Dirac quantization condition (8.31) [cf. Ref. 69].

In the present chapter we have constructed the supersymmetric generalization of the fibrations of S^3 as discussed above. Let us here briefly examine the corresponding mathematical structure. The supersymmetric version $SU(2)_x$ of SU(2) is defined by letting the group parameters become superfields as indicated by the Equation (9.25). Let

$$U(1)_x = \{\exp(i\sigma_3 \alpha)\} , \qquad (9.51)$$

where α is an even Grassmann variable. $U(1)_x$ has a righthand-
ed action on $SU(2)_x$, i.e.

$$s_x \rightarrow s_x \exp(i\sigma_3\alpha) \tag{9.52}$$

The projection map (9.46) can therefore be generalized to

$$s_x \rightarrow s_x \sigma_3 s_x^\dagger = \sigma_a \hat{x}_{xa} \tag{9.53}$$

where the image of the map (cf. (9.30)) is the supersymmetric
version S^2_x of the two-sphere S^2. The bundle which describes
the spinning charge-monopole system is then

$$L_{M_x} = SU(2)_x/Z_M \tag{9.54}$$

As will be shown in section 9.5 the Dirac quantization condition (8.31) is fullfilled also in this case, i.e. we choose
$M = |2n|$.

The $U(1)_x$ gauge transformation will induce a transformation
on the ξ -variables defined in the Equation (9.26). In fact,
if we write

$$\exp(i\sigma_3\gamma(t,\theta)) = (1 + \theta\sigma_3\beta(t))\exp(i\sigma_3\alpha(t)) \tag{9.55}$$

then by the projection map (9.31) and (9.26) ξ will transform
according to

$$\xi(t) \rightarrow \xi(t) + \beta(t)\ddot{x}(t) , \tag{9.56}$$

i.e. ξ undergoes a translation parallel to $\hat{x}$. Since X_x is gauge
invariant, it determines ξ only up to a transformation (9.56).
Equation (9.33) is consistact with this observation.

9.4. Supersymmetric Yang-Mills-Particles

In the present section we will combine the description
of Yang-Mills particles (Cf. Chapter 6) with the supersymmetry
discussed above. For simplicity, we will restrict ourselves

to nonrelativistic particles, but the discussion can be easily
generalized to the relativistic case [71].

The free part of the Lagrangian will again be given by
(9.19). We now extend the minimal coupling preseription (9.22)
to the non-Abelian case, where the Yang-Mills vector potential
is a matrix (cf. Eq.(6.10)). The supersymmetric generalization
of the Lagrangian (6.9) is therefore

$$L_{\mathbf{x}} = L_{0\mathbf{x}} + L_{I\mathbf{x}} \ , \tag{9.57}$$

where $L_{0\mathbf{x}}$ is given by (9.19) and the minimal coupling term
$L_{I\mathbf{x}}$ is

$$L_{I\mathbf{x}}(t,\theta) = \text{Tr}(Ks_{\mathbf{x}}^{\dagger}(t,\theta)D_{t,\theta}s_{\mathbf{x}}(t,\theta)) \tag{9.58}$$

Here we have generalized the covariant derivative appearing
in Equation (6.10) to

$$D_{t,\theta} = d_{\theta} - ie(d_{\theta}X_a(t,\theta))A_a(X(t,\theta)) \tag{9.59}$$

Next we expand the dynamical group element $s_{\mathbf{x}}(t,\theta)$ (cf. (9.26))

$$s_{\mathbf{x}}(t,\theta) = (1 + \theta\xi)s(t), \quad \xi = \xi_a T(a) \tag{9.60}$$

and the supercoordinate $\vec{X}(t,\theta)$ according (9.13), and we in-
tegrate equation (9.57) with respect to θ, i.e.

$$L(t) = \int d\theta L_{\mathbf{x}}(t,\theta) = L_0(t) + L_I(t) \tag{9.61}$$

The result is that L_0 is given by (9.1) and

$$L_I = - \ iTr(Ks^{\dagger}D_t s) - ie/mTr(If_a f_b \partial_b A_a) -$$
$$- Tr(\ I\xi\xi \ + e[I,\xi]A_a f_a/\sqrt{m}) \tag{9.62}$$

where I is given by (6.13) and D_t is the same as in (6.10).
Since the Lagrangian (9.61) does not contain time derivatives
of the dynamical variable ξ, ξ plays the role of an auxiliary

field (see e.g. Ref. 56). The ξ-variable in the Lagrangian is necessary in order that succesive supersymmetric transformations, induced by the translations

$$t \rightarrow t + i\varepsilon\theta, \quad \theta \rightarrow \theta - \varepsilon \qquad (9.63)$$

close with out the use of the equations of motion (see e.g. Ref. 57). We are allowed to substitute the equation of motion for ξ i.e.

$$[\xi, I] = - \frac{e}{\sqrt{m}} f_a [A_a, I] \qquad (9.64)$$

back into the Lagrangian (9.62). Equation (9.64) leads to

$$Tr(I\xi\xi) = \frac{e^2}{2m} f_a f_b Tr(I[A_a, A_b]) \qquad (9.65)$$

After substituting (9.64) and (9.65) into (9.62), we find

$$L_I = -iTr(Ks^\dagger D_t s) - \frac{e}{m} \vec{S} \cdot Tr(I\vec{B}) \qquad (9.66)$$

where $\vec{B}$ is the non-Abelian magnetic field strength.

Concerning the equations of motion derived from the Lagrangian (9.66) (or (9.62)), we notice that the spin precession equation (9.38) will be modified according to

$$\dot{S}_a = \frac{e}{m} \varepsilon_{abc} B_b^\alpha I_\alpha S_c \qquad (9.67)$$

Thus the gyromagnetic ratio is two, as expected.

9.5. Canonical Formulation and Quantization of Pseudo-Classical Systems.

In deriving the comonical formalism for the preceding systems, we follow the methods used in Chapter 8. For treating the fermionic variables, we shall apply the methods of Ref. [53], which are as follows:

Let χ_a denote the momenta conjugate to f_a. If C and D are any anticommuting variables, then PB is defined according to

$$\{C,D\} = - \left(\frac{\partial C}{\partial f_a} \frac{\partial D}{\partial x_a} + \frac{\partial C}{\partial x_a} \frac{\partial D}{\partial f_a} \right) \tag{9.68}$$

Hence,

$$\{f_a, f_b\} = \{x_a, x_b\} = 0$$

$$\{f_a, x_b\} = -\delta_{ab} \tag{9.69}$$

The remaining PB's are defined in the usual way.

For the nonrelativistic particle interacting with a magnetic field (Cf. Equations (9.1) and (9.23)),

$$x_a = \frac{\partial L}{\partial \dot{f}_a} = -\frac{i}{2} f_a \tag{9.70}$$

Thus we obtain the primary constraints

$$\zeta_a = x_a + \frac{i}{2} f_a \tag{9.71}$$

The Hamiltonian is

$$H = \frac{1}{2m} \left(P_a + eA_a(x) \right)^2 + \frac{e}{m} \vec{S} \cdot \vec{B} + \lambda_a \zeta_a , \tag{9.72}$$

where λ_a's, are Lagrange multipliers. The requirement that $\{\zeta_a, H\} \approx 0$ determines the λ_a's, i.e.

$$\lambda_a = \frac{e}{m} \epsilon_{abc} B_b f_c , \tag{9.73}$$

and thus leads to no secondary constraints.

The constraints ζ_a are second class, since

$$\{\zeta_a, \zeta_b\} = -i\delta_{ab} \tag{9.74}$$

They many be eliminated by introducing the DB's [53]

$$\{f_a, f_b\}^* = -i\delta_{ab}$$

$$\{f_a, x_b\}^* = -\frac{1}{2}\delta_{ab} \tag{9.75}$$

$$\{x_a, x_b\}^* = \frac{i}{4} \delta_{ab}$$

The DB's which involve x_a or p_a are all equal to the corresponding PB's. Thus we can replace PB's by DB's and then eliminate x_a via equation (9.70).

The generator of the supersymmetry transformation on the phase space variables is

$$Q = \frac{1}{\sqrt{m}} f_a (p_a + eA_a(x)) \quad , \tag{9.77}$$

since

$$\{f_a, Q\}^* = -i/\sqrt{m} (p_a + eA_a(x))$$
$$\{x_a, Q\}^* = 1/\sqrt{m} \, f_a \tag{9.78}$$

(compare with eq. (9.4)) Furthermore, the Hamiltonian (9.72) can be expressed by

$$H = \frac{1}{2i} \{Q, Q\}^* \tag{9.79}$$

In passing to the quantum theory we replace the DB's in (9.75) by (-i) times the anti-commutator brackets (and the remaining DB's by (-i) times the commutator brackets). In particular,

$$\left[f_a, f_b \right]_+ = \delta_{ab} \tag{9.80}$$

It is known as a consequence [53] that an IRR of the f_a's is obtained in the quantum theory by the identification

$$\tag{9.81}$$
$$f_a = \sqrt{1/2} \cdot \sigma_a \quad ,$$

σ_a's being the Pauli matricies. Consequently, the spin of the particle is $\frac{1}{2}$. Furthermore, (9.79) becomes

$$H = Q^2 \tag{9.82}$$

For the monopole system described in section 2 we replace the above variables x_a and p_a by r, p_r, s and t_a (p_r and t_a are cononically conjugate to r and s, respectively (cf. section 8.2)). Variables t_a and s satisby the Poission bracket relations (8.1). For this system, in addition to the constraint (9.71) we have (8.20), i.e.

$$\phi = \hat{x}_a t_a - n \approx 0 \tag{9.83}$$

The Hamiltonian is now

$$H = p_r^2/2m + 1/2mr^2(t_a t_a - n^2) + \frac{e}{m} \vec{S} \cdot \vec{B} + \lambda_a \zeta_a + n\phi, \qquad (9.84)$$

λ_a and η being Lagrange multipliers. As before there are no secondary constraints. The constraints ζ_a are once again second class while ϕ is first class. The former are eliminated via the DB's (9.75), while the gauge symmetry generated by the latter is eliminated by working on the reduced phase space, which is coordinatized by r, p_r, t_a and $\hat{x}_a$ [cf. section 8.2].

For the monopole system we can express the supersymmetry generator globally by

$$Q = \frac{1}{\sqrt{m}} (f_a \hat{x}_a p_r - \frac{1}{r} t_a \epsilon_{abc} f_b \hat{x}_c) , \qquad (9.85)$$

as compared with (9.77). After applying the constraint (9.83), we can once again show that Hamiltonian is given by (9.79).

In passing to the quantum theory we again make the identification (9.80), yielding the spin-half particle. The quantization of the remaining variables is the same as in section 8.2. In particular, quantization is possible only if $2n$ = integer.

Next, we take up the canonical quantization of the supersymmetric Yang Mills particle described in section 4. We pick up the discussion with the interaction Lagrangian (9.66), where the auxiliary variables ξ have allready been eliminated. The corresponding phase space for this system is spanned by x_a, s, f_a and the canonically conjugate variables p_a, t_α, χ_a. Variables t_α and s again satisfy the Poission bracket relations (8.7). The bosonic variables are constrained by equation (8.74), i.e.

$$\phi_\alpha = I_\alpha - t_\alpha \approx 0 \quad , \tag{9.86}$$

while the fermionic variables are constrained by (9.71). The Hamiltonian for this system is

$$H = 1/2m \left(p_a - eA_a^\alpha(x)t_\alpha \right)^2 + \frac{e}{m} S_a t_\alpha B_a^\alpha + \lambda_a \zeta_a + \eta_\alpha \phi_\alpha \tag{9.87}$$

The treatment of the bosonic constraints and the fermionic constraints have both been previously discussed (the former in section 8.4). Here the supersymmetry generator is

$$Q = 1/\sqrt{m} (p_a - eA_a^\alpha(x)t_\alpha) f_a \tag{9.88}$$

It can have a non trivial action on the isospin variables when an external field is present

$$\{t_\alpha, Q\}^* = -e/\sqrt{m} \ f_a \ c_{\alpha\beta\gamma} A_a^\beta t_\gamma \tag{9.89}$$

The quantum theory for the above system describes a particle of spin half and isospin which is determined by the value of the constants K_α.

10. Local and Global Lagrangians

In the previous Chapters, we considered systems which admit a global Hamiltonian description. That is, these systems have a globally defined Hamiltonian or energy function, and the corresponding symplectic form (or equivalently, the Poisson bracket) is globally defined. However, these systems do not admit global canonical coordinates. Thus a global Lagrangian cannot be found in terms of the variables which occur in the Hamiltonian description. Now by a theorem of Darboux [72], local canonical coordinates always exist. Thus, locally, the Legendre transform can be made and a Lagrangian can be found. These local Lagrangians are defined on coordinate neighbourhoods and are, in general, not defined globally. In previous Chapters, in effect, we have constructed global Lagrangians from these local ones by introducing additional gauge degrees of freedom, that is, a principal fibre bundle structure.

In this Chapter we give a systematic method for finding the global Lagrangian when the system admits local Lagrangians and a global Hamiltonian description. The analysis presented here is similar to an analysis used in context of geometric quantization.

Three striking results emerge from the analysis: 1) The construction in terms of $U(1)$ fibre bundles works only if, classically, a certain "quantization" is fulfilled. For the system of several charges and monopoles, this result has been proved by Friedman and Sorkin [19]. For that system, the condition is

$$\frac{e_i g_j}{e_k g_l} = \text{a rational number,} \qquad (10.1)$$

where e_i and g_i are electric and magnetic charges. Note that this implies that electric and magnetic charges (and hence their product) are seperately quantized (Take $g_j = g_\ell$ to get the first result. Take $e_i = e_k$ to get the second result.) Note also that (10.1) is weaker than Dirac's result

$$e_i g_j = 2\pi k , \quad k \text{ integer,} \tag{10.2}$$

the proof of which requires quantum mechanics.

2) Once the quantization condition is fulfilled, a global Lagrangian can be found by introducing $U(1)$ gauge degrees of freedom, that is a $U(1)$ fibre bundle. It is interesting that in such a case nothing more involved than a $U(1)$ fibre bundle is required for the construction of the global Lagrangian.

3) Global Lagrangians can be constructed even if the quantization condition is not fulfilled and hence the fibre bundle approach fails. The fibre bundle construction is a special case of this more general construction.

In the proof of these results, we use the language of differential geometry because of its convenience. We have done so sparingly however, so that a reader with a small familiarity with differential geometry can follow the argument.

10.1 The Fibre Bundle Construction

Before discussing the main result, we first recall the proof of a theorem due to Weil [73]. For our purposes, Weil's result can be stated as follows: Let Ω be a closed two-form on Q, i.e.

$$\Omega = \Omega_{ij} \, dx^i \wedge dx^j , \tag{10.3}$$

$$d\Omega = 0 \quad \text{or} \quad \partial_i\Omega_{jk} + \partial_j\Omega_{ki} + \partial_k\Omega_{ij} = 0 \qquad (10.4)$$

Further, for every closed two-surface M in Q, let

$$\int_M \Omega = 2\pi\nu\lambda \quad , \quad \nu = 0, \pm 1, \pm 2, \ldots \qquad (10.5)$$

Here λ is the same for all M and ν is characteristic of M. Then there exists a U(1) bundle E on Q, and a form $\tilde{\Omega}$ on E with the following properties:

1) $\tilde{\Omega}$ is exact, ie. $\tilde{\Omega} = d\Lambda$

Here Λ is a globally defined one-form on E.

2) $\tilde{\Omega}$ is "gauge invariant" and hence projects down to a form on Q.

3) The latter is precisely Ω.

Here by gauge invariance we mean the following: Let ϕ and ϕ' be two sections. (Cf. Chapter 3) from a coordinate neighbourhood in Q to E. Then the pull backs $\phi^*\tilde{\Omega}$ and $\phi'^*\tilde{\Omega}$ are equal. Stated in another way, let $\pi: E \to Q$ be the projection map from the bundle E to the base Q. Then $\tilde{\Omega} = \pi^*\Omega$.

In conventional classical mechanics, where global canonical coordinates exist, the symplectic form

$$dp_i \wedge dq^i \qquad (10.6)$$

is necessarily exact:

$$dp_i \wedge dq^i = d(p_i dq^i) \qquad (10.7)$$

Weil's result gives us conditions under which a non-exact symplectic form can be turned into an exact one. This is accomplished by introducing gauge degrees of freedom. Note in this context the "quantization" of the integrals in (10.5). The origin of the classical quantization condition is this equation.

If $\lambda = 0$, then Ω is exact. We shall assume here after

that $\lambda \neq 0$. We also assume that Q is paracompact. Under this

technical assumption, Q has a contractible covering $\{U_\alpha\}$ by

coordinate nerghbourhoods U_α . In such a covering, each of

the sets $U_\alpha, U_\alpha \cap U_\beta, U_\alpha \cap U_\beta \cap U_\gamma, \ldots$ is either empty or can be

smoothly contracted to a point. The proof of the converse to

the Poincaré lemma $\boxed{72}$ is therefore valid on each of these

sets. It follows from (10.3) that

$$\Omega \mid U_\alpha = d\theta_\alpha \quad , \qquad (10.8)$$

where $\Omega \mid U_\alpha$ is the restriction of Ω to U_α . Also, since

$d (\theta_\alpha - \theta_\beta) = 0$ on $U_\alpha \cap U_\beta$, we have

$$\theta_\alpha - \theta_\beta = df_{\alpha\beta} \qquad on \qquad U_\alpha \cap U_\beta \qquad (10.9)$$

where

$$d (f_{\alpha\beta} + f_{\beta\gamma} + f_{\gamma\alpha}) = 0 \qquad on \quad U_\alpha \cap U_\beta \cap U_\gamma \qquad (10.10)$$

Equation (10.10) states that $f_{\alpha\beta} + f_{\beta\gamma} + f_{\gamma\alpha}$ is

a constant on $U_\alpha \cap U_\beta \cap U_\gamma$. Suppose further that

$$f_{\alpha\beta} + f_{\beta\gamma} + f_{\gamma\alpha} = 2\pi n_{\alpha\beta\gamma} \lambda \quad , \qquad (10.11)$$

where $n_{\alpha\beta\gamma}$ has integer values. Then the map

$$F(f_{\alpha\beta}) = g_{\alpha\beta} = \exp (i f_{\alpha\beta} / \lambda) \qquad (10.12)$$

fulfills the cocycle property

$$g_{\alpha\beta} \, g_{\beta\gamma} \, g_{\gamma\alpha} = 1 \qquad on \quad U_\alpha \cap U_\beta \cap U_\gamma \qquad (10.13)$$

The functions $g_{\alpha\beta}$ are defined on $U_\alpha \cap U_\beta$ and have values in $U(1)$.

Hence they define a $U(1)$ bundle on Q.

It may be shown $\boxed{73}$ that (10.11) is equivalent to

(10.5). Thus with (10.13), we have a $U(1)$ bundle on Q. It is

defined as follows. Let x and x' be the coordinates of the
same point p in $U_\alpha \cap U_\beta$ for the coordinate systems appropriate
to U_α and U_β. Then $(x, h^{(\alpha)})$ and $(x, h^{(\alpha)} g_{\alpha\beta})$ define the same
point in the fibre over p in the bundle space E. Here $h^{(\alpha)} \varepsilon U(1)$.
Such a definition of principal fibre bundles is equivalent
to the definition we gave in Chapter 3.

Let
$$m_\alpha = -i \lambda h^{(\alpha)^{-1}} d h^{(\alpha)} \qquad (10.14)$$

The form m_α is defined on the fibres over U_α in the coordinate
system appropriate to U_α. We have,

$$m_\alpha - m_\beta = i \lambda g_{\alpha\beta}^{-1} d g_{\alpha\beta} = - d f_{\alpha\beta} \qquad (10.15)$$

on $U_\alpha \cap U_\beta$. Comparison of (10.9) and (10.15) shows that

$$\theta_\alpha + m_\alpha = \theta_\beta + m_\beta \qquad (10.16)$$

Thus the one form

$$\theta = \theta_\alpha + m_\alpha \qquad (10.17)$$

is globally defined on E. Further, since

$$d\, m_\alpha = 0 \qquad (10.18)$$

we can write $\Omega = d\theta$ if we regard Ω as a form on E. (More correctly,
it is the form $\tilde{\Omega}$ in the statement of the theorem). The theorem
is thus proved.

In the statement of our result, we regard the Hamiltonian or energy and the symplectic form as defined in terms of
coordinates and velocities (and not in terms of coordinates
and momenta). We define Q to be the configuration space for a
dynamical system. Let $\{U_\alpha\}$ be a contractible covering of Q
(assumed paracompact) by coordinate neighbourhoods U_α and TU_α
be the tangent bundle (the space of coordinates and velocities)
over U_α. Suppose now that the following is true: 1) the dynamical system admits local Lagrangians $L^{(\alpha)}$ defined on TU_α.

2) the energy function H is defined globally on $TQ = \bigcup_\alpha TU_\alpha$. In local coordinates, this means

$$\frac{\partial L^{(\alpha)}}{\partial \dot{x}_i} \dot{x}_i - L^{(\alpha)} = \frac{\partial L^{(\beta)}}{\partial \dot{x}_i} \dot{x}_i - L^{(\beta)} \qquad (10.19)$$

on $TU_\alpha \cap TU_\beta$ (assumed not null). 3) The symplectic form ω exists globally, that is,

$$d\left[\frac{\partial L^{(\alpha)}}{\partial \dot{x}_i} dx_i\right] = d\left[\frac{\partial L^{(\beta)}}{\partial \dot{x}_i} dx_i\right] \qquad (10.20)$$

on $TU_\alpha \cap TU_\beta$. 4) The integral of ω over any closed two dimensional surface M in Q fulfills an analogue of (10.5):

$$\int_M \omega = 2\pi\nu\lambda \quad , \qquad \nu = 0, \pm 1, \pm 2, \ldots \qquad (10.21)$$

(Here λ is the same for all TQ and ν is characteristic of TQ.) Then there exists a U(1) bundle E on Q and a global Lagrangian on TE for this system.

Both assumptions 2) and 3) are necessary conditions for the existence of a Hamiltonian description. A system of charges and monopoles fulfills these conditions. Condition 4) is surprising in a classical context since it "quantizes" certain integrals of ω . We shall show that for a system of charges and monopoles, it coincides with the Friedman-Sorkin condition mentioned previously.

To prove our result we can proceed as follows. If $\psi_\alpha = \frac{\partial L^{(\alpha)}}{\partial \dot{x}_i} dx_i$ then by (10.20),

$$d(\psi_\alpha - \psi_\beta) = d\left[\frac{\partial}{\partial x_i}(L^{(\alpha)} - L^{(\beta)}) dx_i\right] = 0 \qquad (10.22)$$

Hence, $\psi_\alpha - \psi_\beta$ can be regarded as a closed one form on $U_\alpha \cap U_\beta$. Since $U_\alpha \cap U_\beta$ is contractible,

$$\psi_\alpha - \psi_\beta = df_{\alpha\beta} \qquad \text{on } U_\alpha \cap U_\beta \qquad , \qquad (10.23)$$

where $f_{\alpha\beta}$ fulfills (10.11) by (10.21). As before, we can construct a U(1) bundle E on Q and forms m_α with the property (10.15). Hence the form χ defined by

$$\chi = \psi_\alpha + m_\alpha \qquad (10.24)$$

exists globally on E and

$$d\chi = \omega \qquad (10.25)$$

(or more precisely $d\chi = \pi^* \omega$ where π is the projection $\pi : E \to Q$)

Now by the energy condition (10.19),

$$L^{(\alpha)} - L^{(\beta)} = \frac{df_{\alpha\beta}}{dx_i} \dot{x}_i \qquad (10.26)$$

Thus the Lagrangian

$$\tilde{L} = L^{(\alpha)} - i\lambda \, h^{(\alpha)-1} \frac{dh^{(\alpha)}}{dt} \qquad (10.27)$$

is globally defined on TE. Since the last term is (locally) the time derivative of a function, $L^{(\alpha)}$ and $L^{(\beta)}$ also give the same equations of motion. The result is thus proved.

Let us understand the result in terms of the charge-monopole system. If

$$\{ \xi^i, \xi^j \} = \omega^{ij}(\xi) \qquad (10.28)$$

are the PB's for a system of coordinates $\xi = (\xi^1, \xi^2, \ldots \xi^{2n})$ for the phase space, the symplectic form is

$$\omega = \omega_{ij} \, d\xi^i \wedge d\xi^j, \qquad (10.29)$$

where

$$\omega_{ij} \, \omega^{jk} = \delta_i^k \qquad (10.30)$$

For a system of one charge and one monopole, the PB's are given in chapter 4. They imply that

$$\omega = 2mdv_i \wedge dx_i - F_{ij}(x) \, dx_i \wedge dx_j \qquad (10.31)$$

where F_{ij} is the magnetic field

$$F_{ij}(x) = n \frac{\varepsilon_{ijk} x_k}{r^3} \qquad (10.32)$$

If M is a closed surface in Q not enclosing the monopole it follows by Stokes theorem that $\int_M \omega = 0$. If M is a 2-sphere S^2 (with outward orientation) which encloses the monopole we get $\int_M \omega = -4\pi n$. Multiple integrations over S^2 with different orientations effectively correspond to different M. Thus, in general,

$$\int_M \omega = 4\pi n \nu_n$$

(10.33)

where ν_n is an integer.

In comparing with eq. (10.21) we may set $\nu = \nu_n$ and $\lambda = 2n$. Consequently, the requirement (10.21) imposes no restrictions on the system (and only defines λ). This, however, is not the case when more than one monopole is present.

If an additional monopole is introduced to the above system we must add

$$- m \, \varepsilon_{ijk} \frac{x'_k}{r'^3} \, dx'_i \wedge dx'_j$$

(10.34)

to (10.32). Here x' corresponds to the distance between the electric charge and the additional monopole. Now

$$\int_M \omega = 4\pi n \nu_n + 4\pi m \nu_m,$$

(10.35)

where ν_n and ν_m are integers. Consequently, (10.21) implies that

$$\lambda \nu = 2n \nu_n + 2m \nu_m$$

(10.36)

Since (10.36) holds for any M, we can choose it such that $\nu_m = 0$. It then follows that λ equals n times a rational number. Similiarly, by choosing M such that $\nu_n = 0$, we can conclude that λ equals m times a rational number. But then

$$\frac{n}{m} = \text{a rational number}$$

(10.37)

which is consistent with (9.1). (Only here $e_i = e_k$).

The following brief remarks about the consequences of discarding the global energy condition (10.19) may be of interest. If this condition is abandoned, the global nature of the symplectic form [Eq. (10.29)] implies only that

$$L^{(\alpha)} - L^{(\beta)} = \frac{\partial f_{\alpha\beta}}{\partial x_i} \dot{x}_i + \rho_{\alpha\beta} \tag{10.38}$$

where $\rho_{\alpha\beta}$ does not depend on $\dot{x}_i$. Further,

a) $\rho_{\alpha\beta} = -\rho_{\beta\alpha}$ and $\rho_{\alpha\beta} + \rho_{\beta\gamma} + \rho_{\gamma\alpha} = 0$

b) Since $L^{(\alpha)}$ and $L^{(\beta)}$ give the same equations of motion (on $TU_\alpha \cap TU_\beta$), the $\rho_{\alpha\beta}$'s actually are constants and hence are globally defined.

Now let $\{\phi_\alpha\}$ be a partition of unity subordinated to the covering $\{U_\alpha\}$:

$$\text{Supp } \phi_\alpha = U_\alpha \quad , \qquad \phi_\alpha \geqslant 0 \ , \quad \sum_\alpha \phi_\alpha = 1 \tag{10.39}$$

Then the globally defined functions

$$\kappa_\alpha = \sum_\lambda \rho_{\alpha\lambda} \phi_\lambda \tag{10.40}$$

have the property

$$\kappa_\alpha - \kappa_\beta = \rho_{\alpha\beta} \tag{10.41}$$

in view of a).

Thus

$$\hat{L}^{(\alpha)} = L^{(\alpha)} - \kappa_\alpha \tag{10.42}$$

fulfill an equation of the form (10.26) and

$$\hat{L} = \hat{L}^{(\alpha)} - i \lambda h^{(\alpha)-1} \frac{dh^{(\alpha)}}{dt} \tag{10.43}$$

is globally defined on TE. Also ω has the usual relation to $\hat{L}$. However, since κ_α can depend on x, $\hat{L}$ may not give the original equations of motion.

10.2. Global Lagrangians without the Quantization Condition

The discussion which follows is taken from Ref. [74].

The variational principles which follow often involve the phase space as the space $Q = \{\xi\}$. They are thus often related to Hamilton's variational principle.

We shall discuss Hamiltonian systems. Thus a globally defined Hamiltonian H and a globally defined symplectic form ω [Cf. Eq. (10.29)] are assumed to exist. Further ω is closed and nondegenerate:

$$d\omega = 0 \qquad (10.44)$$
and
$$\det \omega_{ij} \neq 0 \qquad (10.45)$$

The Hamilton equations of motion for this system are

$$\frac{\partial H}{\partial \xi^i} = \omega_{ji} \, \dot{\xi}^j \qquad (10.46)$$

Suppose now that ω is exact. By definition, then, there exists a globally defined one form $f = f_i(\xi) \, d\xi^i$, such that

$$\omega = df \qquad (10.47)$$

The equations of motion in this case follow from the global Lagrangian

$$L = f_i(\xi) \, \dot{\xi}^i - H(\xi) \qquad (10.48)$$

In familiar situations where Q admits global canonical coordinates, we see from (10.46) that the variational principle associated with L is just Hamilton's variational principle.

If ω is not exact as for the charge-monopole system, then a global f does not exist. Thus we have to modify the above procedure for finding L. One such procedure was described in the previous section. We now point out an alternative approach.

The first step in the modification is to change the configuration space from Q to the space of paths PQ over Q. It is defined as follows. Let ξ_0 be a fixed reference point in Q. (This point may be chosen at will.) Then a point of PQ is a path γ from ξ_0 to some point ξ :

$$\gamma \in \{ \gamma(\sigma) \mid 0 \leqslant \sigma \leqslant 1, \ \gamma(0) = \xi_0 , \ \gamma(1) = \xi \} \tag{10.49}$$

These paths are defined at a given time. We denote time dependent paths by

$$\gamma(\sigma,t) \qquad [\ \gamma(0,t) = \xi_0 \ ] \tag{10.50}$$

We now show that we can always write an action principle with configuration space as PQ. The procedure of course works also when ω is exact. We illustrate it in this context first.

The Hamiltonian H can first be promoted to a functional $\tilde{H}$ on paths at a given time:

$$\int_0^1 d\sigma \ \tilde{H}[\gamma(\sigma,t)] \ = H[\gamma(1,t)] \tag{10.51}$$

Consider next a family of paths $\gamma(\sigma,t)$ with

$$\gamma(1,t) = \xi(t) \tag{10.52}$$

Thus as σ and t vary, $\gamma(\sigma,t)$ sweeps out a surface Δ in Q with boundary

$$\partial\Delta = \partial\Delta_1 \ \cup \ \partial\Delta_2 \ \cup \ \partial\Delta_3 \tag{10.53}$$

where

$$\partial\Delta_1 = \{ \ \xi(t) \mid t_1 \leqslant t \leqslant t_2 \ \}$$

$$\partial\Delta_2 = \{ \ \gamma(\sigma,t_1) \mid 0 \leqslant \sigma \leqslant 1 \ \} \tag{10.54}$$

$$\partial\Delta_3 = \{ \ \gamma(\sigma,t_2) \mid 0 \leqslant \sigma \leqslant 1 \ \}$$

By applying Stokes' theorem, we can write the action

$$\int_{t_1}^{t_2} dt \ [\ f_i(\xi) \ \dot{\xi}^i - H \ ] \ = \int_{\partial\Delta_1} [\ f_i \ d\xi^i \ - H \ dt]$$

as

$$= \int_\Delta [\ \omega_{ij}[\gamma(\sigma,t)] \ d\gamma^i(\sigma,t) \wedge d\gamma^j(\sigma,t) - \tilde{H}[\gamma(\sigma,t)] \ d\sigma \wedge dt \ ] \ +$$

$$+ \left\{ \int_{\partial\Delta_2} f_i[\gamma(\sigma,t)] d\gamma^i(\sigma,t) - \int_{\partial\Delta_1} f_i[\gamma(\sigma,t)] \ d\gamma^i(\sigma,t) \right\} \qquad (10.55)$$

Since we shall not vary the initial and final paths $\gamma(\sigma,t_1)$ and $\gamma(\sigma,t_2)$ in the variational principle, the expression in the script bracket will not contribute to the equations of motion. The action on the space of paths PQ can thus be taken to be

$$S = \int_\Delta [\ \omega - \tilde{H} d\sigma \wedge dt \ ] \qquad (10.56)$$

It involves only the symplectic form ω and not the one form f. It appears to define a field theory in one "space" and one time.

The action S was derived in the case that ω was exact. However, it involves only ω and thus is expected to be valid even if ω is not exact. This expectation is correct as may be shown by varying

$$S = \frac{1}{2} \int_\Delta \omega_{ij} \frac{\partial\gamma^i}{\partial\sigma^a} \frac{\partial\gamma^j}{\partial\sigma^b} \varepsilon^{ab} \ d\sigma \wedge dt - \int_{\partial\Delta_1} H dt \qquad (10.57)$$

$$\sigma^0 = t \ , \quad \sigma^1 = \sigma \quad , \quad \varepsilon^{10} = -\varepsilon^{01} = 1$$

We find, upon using $d\omega = 0$ (that is, $\partial_i\omega_{jk} + \partial_j\omega_{ki} + \partial_k\omega_{ij} = 0$) and regrouping terms,

$$\delta \ \frac{1}{2} \int_\Delta \omega_{ij} \frac{\partial\gamma^i}{\partial\sigma^a} \frac{\partial\gamma^j}{\partial\sigma^b} \varepsilon^{ab} \ d\sigma \wedge dt = - \int_\Delta d[\ \omega_{ij} \ d\gamma^i \ \delta\gamma^j \ ]$$

$$= - \int_{\partial\Delta_1} \omega_{ij} \ d\gamma^i \ \delta\gamma^j \quad , \qquad (10.58)$$

since $\delta\gamma^j = 0$ on $\partial\Delta_2 \cup \partial\Delta_3$.

Also

$$\delta \ \int_{\partial\Delta_1} H \ dt = \int_{\partial\Delta_1} \frac{\partial H}{\partial\xi^j} \ \delta\gamma^j \ dt \qquad (10.59)$$

Thus the Hamilton equations of motion [Cf. eq. (10.46)] are recovered.

For a charge-monopole system the preceding technique can be directly applied to the conventional local Lagrangian

$$L = L_0 + L_I \quad , \tag{10.60}$$

$$L_0 = \frac{1}{2} m \, \dot{x}^2 \quad , \qquad L_I = e \, A_i(x) \, \dot{x}_i$$

to find the global Lagrangian. Here,

$$\partial_i A_j - \partial_j A_i = F_{ij} = - \frac{g \, \epsilon_{ijk} \, x_k}{4\pi r^3} \tag{10.61}$$

is the monopole magnetic field. The latter is globally defined but, of course, the potential A_i is not.

The space Q is $\mathbb{R}^3 - \{0\} = S^2 \times \mathbb{R}^1$, where

$$S^2 = \{ \, \hat{x}_i = x_i/\sqrt{x_i x_i} \} \tag{10.62}$$

$$\mathbb{R}^1 = \{ \, r = \sqrt{x_i x_i} \mid x_i x_i > 0 \, \}$$

Let the reference point be $\xi_0 = (1, 0, 0)$. Then the space PQ is the space of paths γ radiating from ξ_0. The globally defined action and Lagrangian are

$$S = \int \tilde{L} \, d\sigma \wedge dt \quad , \qquad \tilde{L} = \tilde{L}_0 + \tilde{L}_I \quad ,$$

$$\int d\sigma \tilde{L}_0 (\sigma, t) = L_0(t) \quad , \tag{10.63}$$

$$\tilde{L}_I(\sigma, t) = - \frac{eg}{8\pi} \, \epsilon_{ijk} \, \hat{\gamma}_i(\sigma, t) \, \frac{\partial \hat{\gamma}_j}{\partial \sigma^a}(\sigma, t) \, \frac{\partial \hat{\gamma}_k}{\partial \sigma^b}(\sigma, t) \, \epsilon^{ab}$$

Here

$$\hat{\gamma}_i(\sigma, t) = \gamma_i(\sigma, t)/\sqrt{\gamma_j(\sigma, t) \, \gamma_j(\sigma, t)} \tag{10.64}$$

and we identify $\gamma(1, t)$ with $x(t)$.

Finally we make contact with the fibre bundle approach as follows. If ω fulfills a quantization condition of the form (10.21), we know that there is a $U(1)$ bundle E over Q on which ω

becomes exact:

$$\omega = d\chi \qquad \text{on } E. \qquad (10.65)$$

The action S can be thought of as defined on PE, the path space for E. Thus we now regard Δ as a surface in E. Now

$$\int_\Delta \omega = \int_{\partial \Delta_1} \chi \qquad (10.66)$$

plus terms which are not varied and may be discarded. In this way, we have a globally defined action on E:

$$\int_{\partial \Delta_1} (\chi - H \, dt) \qquad (10.67)$$

Here, as in the treatment of the kinetic energy term for the charge-monopole system, we regard H as being defined on E.

Note that this procedure does not work without the quantization condition.

The quantization condition allows us to reduce the path space PQ to the $U(1) \times U(1) \times \ldots \times U(1)$ - bundle E over Q [with k factors of $U(1)$] when there are k 2-cycles $S_1, S_2, \ldots, S_k$, such that

$$\int_{S_j} \omega = a_j \text{ and } a_i/a_j \text{ is irrational}$$

For further details, see Ref [74] .

GENERAL REFERENCES

R.Abraham and J.E.Marsden, Foundations of Mechanics (Benjamin, Reading, Massachusetts, 1978).

V.I.Arnold, Mathematical Methods of Classical Mechanics, Graduate Texts in Mathematics 60 (Springer Verlag, New York, 1978).

Y.Choquet-Bruhat, C.Dewitt-Morette and M.Dillard-Bleick, Analysis, Manifolds and Physics (Noth-Holland, Amsterdam, 1977).

N.S.Craige, P.Goddard and W.Nahm (Editors), Monopoles in Quantum Field Theory - Proceedings of the Monopole Meeting, Trieste, December, 1981 (World Scientific, Singapore, 1982).

L.D.Faddeev and A.A.Slavnov, Gauge Fields - Introduction to Quantum Theory (Benjamin/Cummings, London, 1980).

H.Flanders, Differential Forms with Applications to the Physical Sciences (Academic Press, New York, 1968).

H.Georgi, Lie Algebras in Particle Physics (Addison-Wesley, New York, 1982).

R.Gilmore, Lie Groups, Lie Algebras and some of their Applications (Wiley, New York, 1974).

C.Itzykson and J.-B.Zuber, Quantum Field Theory (McGraw-Hill, New York, 1980).

N.Jacobson, Lie Albebras (Interscience Publishers, New York, 1962).

S.Kobayashi and K.Nomizu, Foundations of Differential Geometry, Vol.1 and Vol.2 (John Wiley, New York, 1963).

A.Lichnerowicz, Theories Relativistes de la Graviation et D´Electromagnetisme (Masson et Cie, Paris, 1955).

D.J.Simms and N.M.J.Woodhouse,Lectures on Geometric Quantization, Lecture Notes in Physics $\underline{53}$ (Springer Verlag, Berlin, 1976)

N.Steenrod, The Topology of Fibre Bundles (Princeton University Press, New Jersey, 1951).

E.C.G.Sudarshan and N.Mukunda, Classical Dynamics: A Modern Perspective (Wiley, New York, 1974).

K. Sundermeyer, "Constrained Dynamics", Lecture Notes in Physics, 169 (Springer-Verlag, Berlin 1982).

W.Thirring, Classical Dynamical Systems - A Course in Mathematical Physics, Vol.1 (Springer Verlag, Berlin, 1978).

F.W.Warner, Foundations of Differentiable Manifolds and Lie Groups (Scott, Foresman and Co, New York, 1971).

C. von Westenholz, Differential Forms in Mathematical Physics
(North Holland, Amsterdam, 1978).

B.S. DeWitt, Dynamical Theory of Groups and Fields (Blackie
and Sons, London and Glasgow, 1965).

References

1. See e.g. A.P.Balachandran, A.M.Din, J.S.Nilsson and H.Ruperts-
   berger, Phys.Rev.D16(1977) 1036;
   A.P.Balachandran, A.Stern and G.Trahern, Phys.Rev.D19(1979)
   2416;
   M.Daniel and C.M.Viallet, Rev.Mod.Phys.52(1980) 175.

2. P.A.M.Dirac,"Lectures on Quantum Mechanics", Belfer Graduate
   School of Science Monographs Series No.2 (Yeshiva University,
   New York, 1964);
   E.C.G.Sudarshan and N.Mukunda,"Classical Dynamics: A Modern
   Perspective" (Wiley&Sons, New York, 1974);
   A.Hanson, T.Regge and C.Teitelboim,"Constrained Hamiltonian
   Systems" (Accademia Nazionale dei Lincei, Rome, 1976);
   K. Sundermeyer, "Constrained Dynamics", Lectures Notes
   in Physics, 169 (Springer-Verlag, Berlin 1982).

3. This proof that $Q_\alpha = 0$ is due to P.G.Bergmann (private communi-
   cation).

4. See e.g. S.Abers and B.W.Lee, Phys.Rep.9C(1973) 1.

5. P.A.M.Dirac,"The Principles of Quantum Mechanics" (Oxford
   University Press, Oxford 1958).

6. See e.g. H.Flanders,"Differential Forms with Applications
   to the Physical Sciences" (Academic Press, New York, 1963).

7. P.A.M.Dirac, Proc.Roy.Soc.(London) Ser A133(1931)60.

8. P.A.M.Dirac, Phys.Rev.74(1948) 817.

9.   P.A. Horváthy, J. Math. Phys. 20 (1979) 49.

10.   See e.g. J. D. Talman, "Special Functions - A Group
      Theoretic Approach" (W.A. Benjamin, New York, 1968).

11.   A.P. Balachandran, S. Borchardt and A. Stern, Phys. Rev.
      D17 (1978) 3247.

12.   N. Steenrod, "Topology of Fibre Bundles", Princeton
      University Press, Princeton, New Jersey 1951.

13.   G.H. Thomas, La Rivista del Nuovo Cimento 3 (1980) 1;
      T. Eguchi, P.B. Gilkey and A.J. Hanson, Phys. Rep.
      66C (1980) 213.

14.   Y.N. Gribov, Materials for the 12th LNPI Winter School,
      Vol.1, p.147 (1977); Nucl. Phys. B139 (1978) 2717.

15.   I.M. Singer, Comm. Math. Phys. 60 (1978) 7.

16.   G. Parisi and Wu Yongshi, Scientia Sinica 24 (1981) 483.

17.   For a review, see P. Goddard and D.I. Olive, Rep. Prog.
      Phys. 41 (1978) 91.

18.   A. Trautman, Int. J. Theor. Phys. 16 (1977) 561;
      J. Nowakowski and A. Trautman, J. Math. Phys. 19 (1978) 1100.

19.   J.L. Friedman and R.D. Sorkin, Phys. Rev. D20 (1979) 2511;
       A.P. Balachandran, G. Marmo, B-S. Skagerstam and A. Stern,
       Nucl. Phys. B162 (1980) 385; B164 (1980) 427.

20.   P.A. Horváthy, "Classical Action, the Wu-Yang Phase Factor
       and Prequantization", Marseille preprint (1980).

21.   T.T. Wu and C.N. Yang, Phys. Rev. D14 (1976) 437.

22.   For a Kaluza-Klein approach see J.G. Miller, J. Math Phys.
       17 (1976) 643; D. Maisson and S.J. Orfanidis, Phys. Rev.
       D15 (1976) 3608.

23.   A.P. Balachandran, G. Marmo, A. Stern and B-S. Skagerstam,
       Phys. Lett. 89B (1980) 199.

24.   In the case of electromagnetism,see A. Stern and B-S.
       Skagerstam, Physica Scripta 24 (1981) 493 and references
       cited therein.

25.   V. Bargmann, L. Michel and V.L. Telegdi. Phys. Rev. Lett.
       2 (1959) 435.

26.   M. Mathisson, Acta Phys. Pol. 6 (1937) 163;
       A. Papapetrou, Proc. R. Soc. London 209A (1951) 248.

27.   F.W. Hehl, Phys. Lett. 36A (1971) 225;
       A. Trautman, Bull. Acad. Pol. Sci. 20 (1972) 895;
       S Hojman, Phys. Rev. D18 (1978) 2741;

R. Saldati and S. Zerbini, Lett. Nuovo Cim. $\underline{27}$ (1980) 575;

M.Toller, "Extended test particles in geometric fields",

Trento preprint (1982).

28.   M. Bailyn and S. Ragusa, Phys. Rev. $\underline{D15}$ (1977) 3543

and $\underline{D23}$ (1981) 1258.

29.   H. van Dam and Th. W. Ruijgrok, Physica (Utrecht) $\underline{104A}$

(1980) 281.

30.   H.P. Kunzle, J. Math. Phys. $\underline{13}$ (1972) 739.

31.   See e.g. Wu-Yang Tsai, Phys. Rev. $\underline{D7}$ (1973) 1945.

32.   W.G. Dixon, Nuovo Cimento $\underline{38}$ (1965) 1616

See also Ch. Duval, Ann. Inst. Henri Poincaré $\underline{25}$ (1976) 345.

33.   R. Utiyama, Phys. Rev. $\underline{101}$ (1956) 1597;

T.W.B. Kibble, J. Math. Phys. $\underline{2}$ (1961) 212;

F.W. Hehl, P. van der Herde and G.D. Kerlik, Rev. Mod.

Phys. $\underline{48}$ (1976) 393;

Y. Ne´eman and T. Regge, Riv. Nuovo Cimento $\underline{1}$ (1978) and

references cited therein.

34.   See e.g. S. Weinberg, Gravitation and Cosmology (Wiley and

Sons, N.Y. 1972) and References 26 and 27.

35.   M.A. Naimark, Linear Representations of the Lorentz Group

(Pergamon Press, 1964).

36.   W. Pauli, Collected Scientific Papers, Vol.II, p.608

       (Interscience Publishers, 1964);

       A.S. Wightman, Rev. Mod. Phys. **34** (1962) 845;

       W.O. Amerein, Helv. Phys. Acta **42** (1969) 149.

37.   A.P. Balachandran, T.R. Govindarajan and B. Vijaylakshmi,

       Phys. Rev. **D18** (1978) 1950.

38.   S.K. Wong, Nuovo Cimento **65A**, 689 (1970);

       S.Sternberg, Proc. Nat. Acad. Sci. **74** (1977) 5253;

       R. Giachetti, R. Ricci and E. Sorace, J. Math. Phys. **22**

       (1981) 1703;

       Ch. Duval and P. Horvathy, Ann. Phys. (N.Y.) **142** (1981) 10.

39.   For an alternative approach which involves the use of

       anticommuting variables in the classical Lagrangian,

       see A.P. Balachandran, P. Salomonson, B-S. Skagerstam

       and J-O. Winnberg, Phys. rev. **D15**, 2308 (1977);

       A. Barducci, R. Casalbuoni and L. Lusanna, Nucl. Phys.

       B124, 93 (1977). The approach given here yields extra

       equations of motion not originally present in the Wong

       equations.

40.   See e.g. L. Michel, Rev. Mod. Phys. **52**, 617 (1980) and

       references cited therein;

       L. Michel and L.A. Radicati, Ann. Inst. H. Poincaré **18**,

       185 (1973).

41.   A.M. Polyakov, Zh. Eksp. Teor. Fiz. Pis'ma, Red. 20, 430
      (1974) JETP Lett. 20, 194 (1974) ;
      G. 't Hooft, Nucl. Phys. B79, 276 (1974).

42.   J. Schechter, Phys. Rev. D14, 524 (1976). The motion of
      the particle at short distances from the center of the
      't Hooft-Polyakov monopole was investigated by
      A. Stern, Phys. Rev. D15, 3672 (1977)

43.   Th. Kaluza, Sitz. Preus. Akad. Wiss. (1921) 966
      O.Klein, Z. Phys. 37 (1926) 895.

44.   B.S. DeWitt, Dynamical Theory of Groups and Fields
      (Blackie and Son, London and Glasgow, 1965).

45.   See e.g. Y.M. Cho and P.G.O. Freund, Phys. Rev. D12
      (1975) 1711;
      Y.M. Cho, J. Math. Phys. 16 (1975) 1711;
      J. Scherk and J.H. Schwarz, Nucl. Phys. B153 (1979) 61.
      For a recent discussion and for additional references
      see e.g. A. Salam and J. Strathdee, Ann. Phys. (N.Y.)
      141 (1982) 316.

46.   A.P. Balachandran, A. Stern and B-S. Skagerstam,
      Phys. Rev. D20 (1979) 439.

47.   N.K. Nielsen, Nucl. Phys. B167 (1980) 249.

48.   N. Mukunda, H. van Dam and L.C. Biedenharn, Phys. Rev.
      D22 (1980) 1938, ibid. D23 (1981) 1451.

49.   D.V. Volkov and S.V. Peletminskii, Zh. Eksp. Fiz.
        37 (1959) 170.

50.   I.L. Martin, Proc. Roy. Soc. (London) A251 (1959) 536.

51.   J.R. Klauder, Ann. Phys. (N.Y.) 11 (1960) 123;
        F.A. Berezin and M.S. Marinov, Pis´ma Zh. Eksp. Teor.
        Fiz. 21 (1975) 678; Ann. Phys. (N.Y.) 104 (1977) 336;
        L. Brink, S. Deser, B. Zumino, P. Di Vecchia and P. Howe,
        Phys. Lett. B64 (1976) 435;

        L.Brink, P.Di Vecchia and P.Howe, Nucl.Phys. B118(1977) 76;
        V.D.Gershun and U.I.Tkach, Pišma Zh.Eksp.Teor.Fiz. 29(1979) 9;
        D.V.Volkov and A.I.Pashnev, Theor.Math.Phys. 44(1980) 3;
        M.Henneaux and C.Teitelboim, Ann.Phys.(N.Y.) 143(1982) 127.

52.   C.A.P.Galvao and C.Teitelboim, J.Math.Phys. 21(1980) 1983.

53.   R.Casalbuoni, Nuovo Cimento 33A(1976) 389.

54.   A.Barducci, R.Casalbuoni and L.Lusanna, Nucl.Phys. B124(1977)
        521;
        P.Salomonson, B-S.Skagerstam and J-O.Winnberg, Phys.Rev.
        D16 (1977) 2581.

55.   See e.g. F.A. Berezin and G.I. Kac, Math. USSR Sbornik 11
        (1970) 311;
        L. Corwin, Y. Ne´eman and S. Sternberg, Rev. Mod. Phys.
        47 (1975) 573.

56.   P. Fayet and S. Ferrara, Phys. Rep. <u>32</u> (1977) 25;

        P. Van Nieuwenhuizen, Phys. Rep. <u>68</u> (1981) 189;

        J. Wess, Lectures given at Princeton University (1981);

        S. Weinberg, Lectures given at University of Texas, Austin

        (1982).

57.   See e.g. P. Salomonson, Phys. Rev. <u>D18</u> (1978) 1868.

58.   For the superfield formalism see e.g. A. Salam and J.

        Strathdee, Phys. Rev. <u>D11</u> (1975) 1521; Nucl. Phys. <u>B76</u>

        (1974) 477.

59.   F.A. Berezin, The Method of Second Quantization, Academic

        Press, New York 1966.

60.   E. Witten, Nucl. Phys. <u>B188</u> (1981) 513.

61.   Last reference in Ref. 19.

62.   S. Ferrara and E. Remiddi, Phys. Lett. <u>53B</u> (1974) 347.

63.   P. Salomonson and J.W. Van Holten, Nucl. Phys. <u>B196</u> (1982)

        509;

        J.W. Van Holten in Proceedings of the 2nd European Study

        Conference on the Unification of Fundamental Interactions,

        Erice, 1981 (Ref. TH. 3191- CERN, 1981).

64.   F. Cooper and B. Freedman, Ann. Phys. (N.Y.) (in press).

65.   J.D. Jackson, Classical Electrodymanics (Second edition, John Wiley and Sons, New York, 1975).

66.   P.J. Hilton and S. Wylie, Homology Theory (Cambridge University Press, 1962).

67.   L.H. Ryder, J. Phys. A.: Math. Gen. 13 (1980) 437 (see, however, remarks in the introduction of Ref. 69).

68.   M. Minami, Prog. Theor. Phys. 62 (1979) 1128.

69.   M. Quiro´s, T. Rami´rez Mittelbrunn and E. Rodri´ques, J. Math. Phys. 23 (1982) 873.

70.   See Ref. 12, page 135.

71.   See Ref. 46, paragraph VII.A.

72.   See e.g. V. Von Westenholz, Differential Forms in Mathematical Physics (North- Holland, Amsterdam, 1978).

73.   D.J. Simms and N.M.J. Woodhouse, Lectures on Geometric Quantization (Springer Verlag. Berlin, 1976) p. 121.

74.   F. Zaccaria, E.C.G. Sudarshan, J.S. Nilsson, N. Mukunda, G. Marmo and A.P. Balachandran, Institute of Theoretical Physics preprint 81-11, Göteborg 1981 and Phys. Rev. D (in press).

Lecture Notes in Physics

Editors: J. Ehlers, K. Hepp, R. Kippenhahn,
H. A. Weidenmüller, J. Zittartz

Volume 159
E. Seiler

Gauge Theories as a Problem of Constructive Quantum Field Theory and Statistical Mechanics

1982. V, 192 pages. ISBN 3-540-11559-5

Contents: Introduction. – Lattice Gauge Theories: The Scheme of Lattice Gauge Theories. Fundamental Properties. Expansion Methods. Further Developments. – Continuum Gauge Quantum Field Theories: Approaches to the Construction of Continuum Gauge Quantum Field Theories. Convergence to the Continuum Limit in External or Cutoff Gauge Fields. Removal of All Cutoffs, Verification of Axioms in Two Dimensions. A Framework for Non-local Gauge Invariant Objects. – Appendix: The Geometric Setting of Gauge Theories. – References.

Volume 169
K. Sundermeyer

Constrained Dynamics

with Applications to Yang-Mills Theory, General Relativity, Classical Spin, Dual String Model

1982. IV, 318 pages. ISBN 3-540-11947-7

Contents: Introduction. – Classical Regular Systems. – Classical Singular Systems. – The Reduced Phase-Space. – Quantization of Constrained Systems. – The Electromagnetic Field. – Yang-Mills Theory. – The Relativistic Particle. – The Relativistic String. – Einstein's Theory of Gravitation. – Appendices A-E. – Synopsis. – References.

Volume 174
A. Kadić, D. G. B. Edelen

A Gauge Theory of Dislocations and Disclinations

1983. VII, 290 pages. ISBN 3-540-11977-9

Contents: Historical Remarks and Phenomenology of Defects. – Preliminary Considerations. – The Gauge Theory of Defects. – Linearizations. – Appendix 1: The Lie Algebra of $SO(3) \triangleright T(3)$. Appendix 2: Invariance of L_0 under $SO(3) \triangleright T(3)$. – Appendix 3: Induced Transformations of the Field Variables. – Appendix 4: A Four-Dimensional Formulation of Defects. – Dynamics and Thermodynamics. – References.

Volume 176

Gauge Theory and Gravitation

Proceedings of the International Symposium on Gauge Theory and Gravitation (g & G)
Held at Tezukayama University
Nara, Japan, August 20–24, 1982
Editors: K. Kikkawa, N. Nakanishi

1983. X, 316 pages. ISBN 3-540-11994-9

Contents: Geometrical Aspect of Gauge Theory and Gravitation. – Gauge Theories I. – Quantum Field Theory in Curved Space-Time. – Gauge Theories II. – Special Session. – Gauge Theories III. – Quantum Gravity. – Supersymmetry and Supergravity. – Unified Theories. – New Ideas. – Closing Address. – Contributions by Mail. – Author Index. – List of Participants.

Volume 180

Group Theoretical Methods in Physics

Proceedings of the XIth International Colloquium
Held at Boğaziçi University, Istanbul, Turkey,
August 23–28, 1982
Editors: M. Serdaroğlu, E. İnönü

1983. XI, 569 pages. ISBN 3-540-12291-5

Contents: Group Representations. – Completely Integrable Systems and Group Theory. – Gravity, Supergravity, Supersymmetry. – Crystal Groups and their Representations. – Elementary Particles, Grand-Unification, Gauge Theories. – Symmetry Breaking Group Contraction and Extension and Bifurcation. – Nuclear Physics. – Geometrical Methods in Quantum Mechanics and Field Theory. – Statistical Mechanics. – List of Participants. – Appendix I: Acceptance Speech of Prof. Gel'fand on the Award of the Wigner Medal for 1980. – Appendix II: Presentation of the Wigner Medal for 1982. – Introductory Remarks: A. Böhm. – Tribute to Yuval Ne'eman: L. C. Biedenharn.

Springer-Verlag
Berlin
Heidelberg
New York
Tokyo

Lecture Notes in Physics

Selected Issues from

Lecture Notes in Mathematics